T0038185

SAVING OURSELVES

SOCIETY AND THE ENVIRONMENT

SOCIETY AND THE ENVIRONMENT

The impact of humans on the natural environment is one of the most pressing issues of the twenty-first century. Key topics of concern include mounting natural resource pressures, accelerating environmental degradation, and the rising frequency and intensity of disasters. Governmental and nongovernmental actors have responded to these challenges through increasing environmental action and advocacy, expanding the scope of environmental policy and governance, and encouraging the development of the so-called green economy. Society and the Environment encompasses a range of social science research, aiming to unify perspectives and advance scholarship. Books in the series focus on cutting-edge global issues at the nexus of society and the environment.

Series Editors
Dana R. Fisher and Evan Schofer
Co-Founding Editor: Lori Peek

■ ■ ■

Reforesting the Earth: The Human Drivers of Forest Conservation, Restoration, and Expansion, Thomas K. Rudel
Underwater: Loss, Flood Insurance, and the Moral Economy of Climate Change in the United States, Rebecca Elliott
Super Polluters: Tackling the World's Largest Sources of Climate-Disrupting Emissions, Don Grant, Andrew Jorgenson, and Wesley Longhofer

SAVING OURSELVES

From Climate Shocks to Climate Action

DANA R. FISHER

Columbia University Press

New York

Columbia University Press
Publishers Since 1893
New York Chichester, West Sussex
cup.columbia.edu

Library of Congress Cataloging-in-Publication Data
Names: Fisher, Dana, 1971– author.
Title: Saving ourselves : from climate shocks to
climate action / Dana R. Fisher.
Description: New York : Columbia University Press, [2024] |
Series: Society and the environment | Includes index.
Identifiers: LCCN 2023042614 (print) | LCCN 2023042615 (ebook) |
ISBN 9780231209304 (hardback) | ISBN 9780231557870 (ebook)
Subjects: LCSH: Environmentalism. | Climatic changes—
Social aspects. | COVID-19 Pandemic, 2020—Influence.
Classification: LCC GE195 .F57 2024 (print) |
LCC GE195 (ebook) | DDC 363.7/06—dc23/eng/20231130
LC record available at https://lccn.loc.gov/2023042614
LC ebook record available at https://lccn.loc.gov/2023042615

Printed in the United States of America

Cover design: Noah Arlow

This book is dedicated to all the people who
will rise up with the seas:
I hope to see as many of you as possible on
the other side.

CONTENTS

ACKNOWLEDGMENTS

BECAUSE THIS book is an agglomeration of various projects that I have undertaken over the past twenty-five years, there are many people and institutions to recognize.

The project benefited from a series of research grants that supported my studying climate politics and climate activism in the United States and internationally: The William and Flora Hewlett Foundation, the John D. and Catherine T. MacArthur Foundation, the U.S. National Science Foundation (NSF), the Swiss Federal Institute of Aquatic Science and Technology (EAWAG), the Oeschger Center for Climate Change Research at the University of Bern, and the Norwegian Research Council have all supported research that is included in this book.

The data presented in this book include those collected from policy actors, as well as extended interview data collected from numerous organizers and activists who gave their time generously to speak with me while they were working as both insiders and outsiders in response to the climate crisis.

Data analysis and early writing was conducted during my time as Distinguished Visiting Scholar at the John W. Kluge Center of the Library of Congress in 2022. It was such a treat to have this time at the Library of Congress to work on the project. I will never forget my time coding interview transcripts in the most beautiful reading room in the world! Because I had used the Library of Congress as my informal Capitol Hill base while I was conducting the U.S. component of my dissertation research twenty-five years ago, it was an extra-special treat to get to have this time and an official desk to use at the Kluge Center.

The book also benefited from informal conversations with colleagues at the Brookings Institution, where I have served as Non-Resident Senior Fellow for the past two years, along with many conversations with Andrew Jorgenson, Lorien Jasny, and Erika Svendsen, all of whom provided their unique perspectives and a sounding board for my ideas as they developed. I could not wish for a better set of colleagues and collaborators.

Thanks are also due to Adam Met for the many conversations we had about on-ramps to engagement and climate activism as I was getting started on the writing, as well as to Lyle Andrews, for taking time out of his busy schedule in summer 2023 to give me input on an almost-final draft while I was finishing my editing. Finally, thanks to Peloton for keeping me sane . . . I discovered you when the world shut down and you kept me going through the pandemic and beyond.

I also want to thank the students from my spring 2023 Global Activism class at the University of Maryland. To Mohammed,

Emmanuel, Viktoriia, Veyli, and Jessica: thank you for talking through my ideas and how they connect to the social movements literature, as well as for sharing your personal experiences as leaders and activists in various global movements. I consider myself extremely lucky to have had you all in my class. You are an inspiration, and you give me hope for the future!

Once again, I want to thank my editor, Eric Schwartz, who was open to thinking about the climate crisis in a very different way and willing to talk me through the dark spaces. He was a supporter of the ideas that turned into this book project from the very beginning. The book is better thanks to his input and the great work of his colleagues at Columbia University Press.

Finally, I want to thank my children: Margot and Conrad. Hopefully, you will look back on my time hiding in my office editing and writing and talking apocalyptic optimism over meals fondly. I sincerely hope that when you are my age, you will be able to think about this period of writing and stressing and feel like it was a good use of my time. Last, but definitely not least, thanks to my Aaron: I couldn't do any of this without you and your support. I couldn't wish for a better coparent, partner, and friend!!

SAVING OURSELVES

Chapter One

NO ONE ELSE IS GOING TO SAVE US

UNDERSTANDING THE SOCIAL SIDE OF THE CLIMATE CRISIS

SOCIAL SCIENTISTS are not supposed to prognosticate. We theorize. We hypothesize. We collect data. We analyze. Then we adjust our expectations based on what we find, and we do it all over again. I have followed this process in my work for over twenty-five years, studying climate politics, activism, civic participation, and democracy. Although some of my findings have informed decision making and contributed to the outcomes that I was studying, my intention was not to prescribe behavior or influence policy.

This book is a substantial departure from all that previous work. I am motivated by the fact that, during my years studying climate policymaking and activism, the problems that I have been studying have not gotten any better. In reality, they have gotten a lot worse, and we are running out of time.

In this book, I synthesize various types of data that I have collected and analyzed over my career to provide evidence about how society is responding to (and exacerbating) the climate crisis. Instead of going back to the drawing board to continue the process of knowledge development, this time, my predictions about the future will lead to suggestions about what we need to do. My hope is that this book will play a role in preparing us all for what is coming and help to minimize the human suffering that will be a likely side effect. If I do it right, my predictions will scare you, but they should also give you hope.

I came of age in the promise and then frustration of the Clinton administration (anyone remember the failed British thermal unit [BTU] tax?). Since then, I have experienced the whiplash of various political administrations and their priorities. I watched firsthand as the Trump administration dismantled large sections of our federal bureaucracy (including parts that had been expanded to address climate change). I have seen how elected officials and political offices shift their positions in response to the outcomes of elections, terrorist attacks, economic downturns, energy prices, and war. I have also observed how specific policy instruments come in and out of political favor without ever being implemented.

Consequently, scientists like me have gone from encouraging politicians to pass climate policies out of an abundance of caution,[1] to trying to figure out what else we can do to get policymakers to act in response to the dire findings from our research with the urgency that it requires. No longer is climate change a possibility to be avoided; it is an inevitability that we must all work together to curtail. As a result, it has become increasingly common for scientists to take to the streets in protest, in some cases even blocking the entrances to buildings while they scream to get policymakers to take their findings seriously.[2]

LIVING IN A WORLD OF CLIMATE SHOCKS

As scientists struggle to get the word out and be heard, people around the world are experiencing the effects of climate change firsthand. NASA'S Goddard Institute for Space Studies (GISS) reported that the world had warmed by at least 1.1°C (1.9°F) since 1880 (and that estimate did not include data from 2023)[3] Now, people around the globe are experiencing *climate shocks*. Climate shocks are deviations from normal environmental patterns in the form of droughts, floods, heat waves, or other extreme events that have been exacerbated by climate change.[4]

In early 2023, UN Secretary-General António Guterres spoke at a conference that was held in the wake of the devastating floods that left more than 9 million people homeless in Pakistan. His address discussed the feasibility of keeping global warming within the 1.5°C goal of the Paris Agreement. "We are on the road to climate ruin . . . The 1.5 degree warming limit—universally agreed as the only way to safeguard our planet and our future—is on the verge of collapse . . . Moving as we are moving, we will get to 2.8 degrees of increased temperature." Guterres continued by stressing the well-known consequences of the warming that is coming: "As global greenhouse gas emissions continue to rise, we are all in danger . . . Today it's Pakistan. Tomorrow it could be in your country."[5]

In 2021, about one in three Americans were estimated to have personally suffered from climate change–exacerbated weather disasters.[6] Climate shocks were even more widespread in 2022 when heat waves, droughts, and wildfires ravaged numerous parts of the globe, along with hurricanes and tropical cyclones that were intensified by waters warmed by climate change.[7] At the end of 2022, the *Guardian* announced that it had been a "'devastating' year on the frontline of the climate crisis."[8] These trends continued in 2023, with the *New York Times* reporting in May, "Climate Shocks Are Making Parts of America Uninsurable."[9] These trends were observed as the world braced for the return of El Niño, which was expected to "push toward levels of global warming that climate scientists have warned could be devastating."[10] Even before the onset of El Niño, however, the World Meteorological Organization (WMO) reported that in 2023, the world "will breach the 1.5°C level on a temporary basis with increasing frequency."[11]

While a high percentage of the world experiences climate shocks, we continue on a path that is guaranteed to lead to even more warming. At the onset of the war in Ukraine in February 2022, there were numerous discussions about how the conflict could provide the impetus for an energy transformation in Europe that would shift it away from natural gas and oil extracted in Russia to cleaner and more renewable energy sources.[12] Unfortunately, the transition did not come. Instead, most countries leaned back into fossil fuels. Germany, for example, shifted away from its plans to phase out coal by 2030 and burned substantially more coal to generate electricity in 2022.[13] In 2023, the country gained media attention for destroying a rural village to expand a coal mine.[14] As the findings from the most recent Intergovernmental Panel on Climate Change (IPCC) report make abundantly clear, this path leads to more of the world experiencing climate shocks that are more severe, come more frequently, and last longer.[15]

LEARNING FROM THE PANDEMIC

In this book, I build on insights from society's responses to COVID-19, as well as our recovery from it. In March 2020, the COVID-19 pandemic shut down the world. As populations around the globe hid inside their homes, the environmental benefits of society's responses to this global pandemic were substantial.[16] The threat of the new disease brought about a rapid societal transformation that included significant limitations on mobility and expansions of social welfare. At the same time, an unintended and additional benefit of these measures to limit the spread of the disease was social change that put society on a much more sustainable trajectory in terms of its emissions of harmful greenhouse gases (GHGs). As most of the global population experienced lockdowns and/or travel restrictions within and beyond their borders, energy consumption went down substantially.[17] Global, national, and local responses to the pandemic were in stark contrast to the slow and ineffective ways that the world has responded to the climate crisis. As youth climate activist Greta Thunberg shrewdly observed early in the pandemic: "The coronavirus is a terrible event . . . But it also shows one thing: That once we are in a crisis, we can act to do something quickly, act fast . . . we can act fast and change our habits and treat a crisis like a crisis."[18]

Looking at social responses to the COVID-19 pandemic can help us understand what it will take to get to meaningful climate action. In this book, I apply the framework of the AnthroShift to assess how transformational social change is most likely to emerge and consider what is the most reasonable pathway out of the climate crisis.

THE ANTHROSHIFT AND COVID-19

Social responses to the pandemic and its environmental side effects are consistent with the notion of AnthroShift. In 2019, Andrew Jorgenson and I introduced this new social theory that explains how social actors are reconfigured in the aftermath of widespread perceptions and experiences of risk.[19] When risk becomes so common that it is felt across society, the interrelations among the main actors—coming from the state, market, and civil society sectors—shift substantially. The notion that risk can drive social change is consistent with the scholarly work on the risk society and reflexive modernization.[20] Perhaps Merryn Ekberg best summarizes how this body of research interprets risk: it expands the "traditional concept of risk understood as the sum of the probability of an adverse event and the magnitude of the consequences, to include the subjective perception of risk, the inter-subjective communication of risk and the social experience of living in a risk environment."[21] This viewpoint is also similar to much of the research on climate risk as well as the growing literature on disaster.[22]

The AnthroShift is like other perspectives that consider risk as a social pivot—when the sense of risk is strong enough, people change their behaviors and push social actors to respond to remediate the risk. During my conversation with the U.S. Deputy Special Envoy for Climate Jonathan Pershing in winter 2022, he explained how his office thinks about the ways that risk can motivate action: "Probably the most salient [way we think of] risk of climate . . . [is] the articulation of that risk as a driver for change. If people understand the risk, we believe they may be more likely to change the direction of their policy."[23] I call this heightened and generalized sense of risk that can motivate social change a *risk pivot*.

Beyond its consideration of the role of risk, the AnthroShift is unique in that it also assumes that social change in response to risk pivots are multidirectional. Society can shift to an orientation of social actors that prioritize environmental protections and the emergence of an "environmental state."[24] However, it can also move back to less environmentally sustainable practices.[25]

While the world became accustomed to the risks associated with COVID-19, for example, individuals resumed more normal levels of travel and social mixing even as the disease continued to spread. As a result, the environmental cobenefits of the responses to the pandemic were short-lived, and many states shifted back to prepandemic policies. This flip-flop in policies and behaviors provides a clear example of the multidirectionality of the AnthroShift. As a result, global GHG emissions quickly returned to their original upward trajectory, with emissions in 2021 overshooting expectations.[26] In their comment in *Nature Reviews Earth & Environment*, Liu and colleagues point out that "rebounds are apparent in most sectors and big emitting nations." The authors go on to note that, although the degree to which the pandemic affected emissions was "unprecedented, such crises and rebounds are not unique."[27]

As GHG emissions resumed their climb, the planet continued to heat up. We witnessed numerous examples of the consequences of a changing climate in the form of extreme weather events around the globe, including hurricanes, droughts, floods, heat waves, wildfires, and much more. These climate shocks are documented in detail in the Sixth Assessment Report from Working Group 2 of the IPCC on impacts, adaptation, and vulnerability, which was published in 2022.[28] In the words of this report's government-approved Summary for Policymakers: "Any further delay in concerted global action will miss the brief, rapidly closing window to secure a liveable future."[29]

FROM COVID-19 TO THE CLIMATE CRISIS

Lessons regarding how the world responded to the pandemic provide important insights into what it will take to motivate the social responses needed to address the climate crisis.[30] As I already mentioned, early responses to the pandemic had the unintended side effect of reducing GHG emissions. These side effects were accidental; mitigating climate change was not directly connected to policy responses to the pandemic. It is also worth pointing out that the social responses to COVID-19 are consistent with the research on disaster, which finds that windows of opportunity open to all sorts of policy innovations in the wake of an event.[31] Outside this example from the early days of the pandemic, however, there is extensive evidence that current efforts to address climate change are woefully inadequate from the state, market, and civil society sectors.[32]

This lack of progress is particularly notable because governments have been working for over thirty years to limit the effects of global warming. Even though the climate regime has been holding regular meetings around the world with the stated goal of reducing concentrations of carbon emissions in the atmosphere and coordinating global progress to mitigate against climate change since 1992, there is ample evidence that we are running in place or even going backwards.[33] Despite numerous efforts to address the climate crisis at multiple levels of governance, policymaking has been ineffective at bringing about the emissions reductions necessary to limit global warming below the 1.5°C threshold identified by the IPCC and codified in the Paris Agreement.[34]

As of late October 2021, for example, the annual *UN Emissions Gap Report* noted that national climate pledges going into the twenty-sixth United Nations Framework Convention on Climate Change Conference of the Parties (COP26) negotiations in Glasgow were still "well above the goals of the Paris climate agreement and would lead to catastrophic changes in the Earth's climate."[35] Since that meeting in Glasgow, progress has not picked up.[36] In fact, the 2022 round of climate negotiations in Egypt was so impotent regarding emissions reduction commitments that the UN secretary-general called for a climate summit to be held in New York City in September 2023 prior to the next formal round of negotiations. In the words of Secretary-General Guterres, the meeting would be "a no-nonsense summit. No exceptions. No compromises. There will be no room for back-sliders, greenwashers, blame-shifters or repackaging of announcements of previous years."[37]

This call for more commitments comes on top of growing concerns about the feasibility of the implementation of existing climate pledges.[38] The UN secretary-general discussed this issue on the occasion of the release of the Sixth Assessment

Report from IPCC Working Group 3 on mitigation in April 2022. In his words, the report documents "'a litany of broken climate promises' by governments and businesses."[39] In short, even when countries do commit to climate goals, they are not following through on these commitments.

While countries vary substantially regarding their institutional makeup,[40] these broken promises are apparent in most nations where adequate climate action continues to be out of reach.[41] In the United States, for example, the country is expected to overshoot its original climate commitments set by President Barack Obama at the COP21 round of the climate negotiations in Paris in 2015. These targets were classified as "insufficient" by the independent scientific team at Climate Action Tracker to keep global warming below 3°C.[42]

In 2021, the Biden administration submitted an updated commitment prior to the COP26 round of negotiations that increases the country's pledge and gets the United States closer to achieving the emissions reduction goals of the Paris Agreement. Finally, after years of failed attempts to pass climate legislation through the U.S. Congress and months of stops and starts, the Inflation Reduction Act, which aims to address climate change along with other issues, passed and was signed into law in August 2022. Although the United States having a federal climate policy that was approved by both the Congress and the president after so long is cause for celebration, this bill only made it to President Biden's desk for signature due to giveaways to the fossil fuel industry that were brokered by West Virginia Senator Joe Manchin.[43] And, even with this new policy, the country is still not on a path to fulfill its climate commitments that would stabilize global warming at the 1.5° threshold set by the Paris Agreement.[44]

Outside the United States, it is not much better. Although many other developed countries have filed Nationally Determined Contributions (NDCs) that indicate plans to reduce their emissions in line with the IPCC's targets, the implementation of policies that achieve these intended goals are few and far between.[45] Since the war in Ukraine began in February 2022, many countries have found meeting their climate goals even more challenging.[46]

While country responses have been insufficient, business efforts have been bipolar. In contrast to the swift and effective global response to ozone depletion, where a technological fix was discovered and companies encouraged governments to implement it,[47] the climate crisis has no silver bullet. To date, companies representing non-carbon-emitting energy sources and technologies continue to butt heads against entrenched business interests that support an economy run on fossil fuels. Recent research has documented the fact that fossil fuel companies have been well aware since the 1970s that burning fossil fuels would lead to a climate crisis.[48] Instead of acting on that information to limit the risk of a crisis, companies buried their findings and misled the public so they could continue to expand their businesses.

At the same time that fossil fuel expansion continues, many companies and governments are investing heavily in the development of technology that will either remove carbon from the atmosphere or reduce solar absorption through geoengineering.[49] Fossil fuel companies are taking advantage of government funding in the United States and abroad to develop direct air capture and carbon capture and storage.[50]

While big oil companies have started to acknowledge the climate issue and propose plans to address climate change, their so-called zero emissions plans do not provide comprehensive strategies. Specifically, these plans intentionally omit reductions in what they call "scope 3 emissions" derived from the actual burning of the natural resources that they are extracting and selling. In other words, these companies have developed plans to reduce the emissions associated with the *extraction* of the fossil fuels, but these plans do not consider the emissions from the *burning* of these fuels because they are sold to other companies and/or countries.[51] Moreover, on the heels of announcements in 2023 about record profits thanks, in part, to the war in Ukraine, many big oil companies announced that they were slowing down the implementation of their climate plans.[52] Given these actions by businesses, there is growing evidence that all commitments should be interpreted with caution.

As a result of this lack of progress in achieving the material goals of the climate regime, many scientists who served as authors for the recent IPCC report have lost confidence that governments will achieve their stated climate goals. In November 2021, *Nature* reported on a survey of scientists who contributed to the most recent assessment report (AR6) of the IPCC. The survey found that many of these scientists "expect to see catastrophic impacts of climate change in their lifetimes."[53] Scientists around the world are reportedly coming to terms with the fact that the world will almost definitely miss its opportunity to meet the Paris climate agreement and cap global warming at 1.5°.[54]

With the lack of meaningful progress to reduce emissions from the state and market sectors over the past few decades, there has been growing pressure from civil society, which includes a range of nonstate actors, such as nongovernmental organizations (NGOs), unions, social movements, individual activists, and even scientists.[55] Climate activism has thus become more common in recent years and has employed an expanding range of tactics that aim to raise awareness of the climate issue and motivate climate action. In September 2019, for example, an estimated 7.6 million people participated in the Global Week for Future actions around the world that were scheduled to coincide with the UN Climate Action Summit.[56]

As the world continues to heat up and policies remain insufficient, a radical flank has grown within the climate movement that is pushing for more confrontational activism.[57] In the lead up to the COP27 round of the climate negotiations in 2022, for example, confrontational civil disobedience became much more common. During that fall, activists blocked streets, stormed football fields, superglued themselves to various places, and threw food on artwork in museums in various countries.[58] As I write, even more days of action and protest are being announced by groups around the world.

To date, however, the effects of climate activism on economic and political actors' behaviors are murky, and even less is known about the actual climate effects of these efforts in terms of how they could bring about emissions reductions.[59] Although climate activism holds promise to pressure state and market actors to respond meaningfully to the climate crisis, we still don't know what sorts of climate activism at what scale are the most effective at leading to reductions in GHG emissions. The research that does analyze how civic action affects carbon emissions and concentrations faces numerous "methodological challenges as the data on the activism, as well as on the connections between the activism and its effects on greenhouse gas emissions are both limited."[60] With the world on a path to more frequent and severe climate shocks that will displace populations around the globe, there is no question that we will see more activism and protest.

GETTING TO CLIMATE ACTION

When I started, I had some pretty rosy expectations about the ways that we could solve the climate crisis (although we didn't call it that back then). All we needed was to discover and invest in a technological silver bullet that would bring about what we called ecological modernization.[61] If we followed the same model used to solve the ozone hole, we could decouple economic development from carbon emissions and industrialize our way to a more ecologically sensitive "third path" of sustainable development.[62]

At that time, the science of climate change had already been well documented by two assessment reports of the IPCC and a third was on the way. While this international panel of climate experts worked tirelessly (and for free) to synthesize the state of knowledge and get the information out to governments and the public, there were concentrated interests with access to power who used every means available to block efforts to reduce

GHG emissions at all scales of governance (including planting experts on the IPCC with the help of specific governments).[63] These interests continue to exercise a disproportionate amount of power over decision makers. Even today, they continue to damper the message of the various IPCC reports and to slow down the progress of the climate regime.[64]

It was only in the most recent Sixth Assessment Report that these vested economic and political interests were finally discussed. However, the report, which was released in 2021–2022, still does not name specific polluters and specific proponents of climate misinformation.[65] Their roles are downplayed even further in the Summary for Policymakers, which is the most read part of the multi-thousand-page reports.

Thirty years into attempts to address climate change, it is abundantly clear that those actors who have amassed their wealth and power through industrialization powered by fossil fuel extraction and combustion have a stranglehold on economic and political power. One such example can be seen in the ways nations are dodging their commitments to reduce their emissions. A number of countries continue to cut emissions within their borders while also permitting fossil fuel expansion with the intent of exporting these resources to other parts of the world.[66] Unless policies change, remember that any emissions from the burning of these fossil fuels fall under the category of scope 3 emissions, so the companies doing the extracting do not include them in whatever climate plans they have (which may or may not be implemented).

The war in Ukraine has exacerbated this process as countries scramble to increase fossil fuel production outside Russia rather than encouraging a faster shift to clean energy sources.[67] One such example is Norway, which has been successful in transitioning its transportation sector away from fossil fuel–powered vehicles. Although the country's domestic plans are on track with the Paris Agreement goals,[68] the country announced that it would offer a record number of oil and gas blocks for exploration in its arctic territories in January 2023.[69] At the same time, countries including Germany and the United Kingdom, which had committed to phasing out fossil fuel extraction, continued to extract and burn coal in response to the energy crisis brought about by the war in Ukraine.

The tentacles of fossil fuel interests spread farther than you may imagine: from the companies that support our addiction to fossil fuels on our roads, in our homes and workplaces, to the elected officials and agency bureaucrats who decide how to invest taxpayer dollars to expand and maintain public infrastructure that tends to prioritize increased reliance on individual transportation (which is fueled by fossil fuels) over public transit. Their influence also reaches into some of the most well-known university institutes with the stated mission of studying (and ostensibly mitigating) climate change.[70] There is no question that these vested interests have played a role in the direction that climate scholarship and inquiry have taken, which is very likely to have contributed to the lack of meaningful political progress on the climate crisis. There is also ample evidence that these interests continue to resist change and push to maintain the status quo at all costs.[71]

As some people cling to the hope that technological innovation and ecological modernization can save us from ourselves, there are limited data to support this perspective.[72] Instead, there is growing resistance to the notion that incremental political change can get us where we need to go. More and more evidence points to the fact that the only way forward involves extensive systemic change,[73] with some scholars calling out the role that capitalism and neoliberal economic policies have played.[74] In 2022, Harvard professor Naomi Oreskes spoke about her work to expose the strategies used by vested interests to obfuscate and distract:[75] "Climate change is an epic market failure. So how can we possibly say the market will fix it, if we don't address the neoliberal understanding of the market?"[76]

With all these influential and expansive drags on climate action, it is no wonder that we have yet to achieve our climate goals. Instead of the systemic change that is woefully needed, we have all been redirected to focus on taking personal steps to make a difference. The only solutions that have received any real political support are ones that push for incremental change that will not fully address the problem. Like many of you, I have tried to do my part: I stopped eating meat, I bought an electric car, and I took out a loan to put solar panels on my house. While I know that these steps can help reduce my personal contributions to the climate crisis and might make me feel better,[77] I am acutely aware that none of them alone or together is enough. In fact, even the United Nations Environment Programme (UNEP) concluded in fall 2022 that "incremental change is 'no longer an option.'"[78]

MAKING THE CASE FOR APOCALYPTIC OPTIMISM

Even though the environmental side effects of the responses to COVID-19 were unintentional, the way society responded to the pandemic provides some guidance regarding what is needed to save ourselves. Risk only motivates an AnthroShift if it surpasses a threshold in terms of its duration and intensity. Without a sustained shock that has tangible consequences in terms of social cost to people and property, the subsequent social change (like we saw during the early days of the pandemic) will be ephemeral. In those cases, the reorientation of social actors will shift back toward the original configuration that maintains a business-as-usual trajectory. In the case of climate change, this trajectory is supported and encouraged by the many actors with vested interests that have consolidated access to resources and power through the fossil fuel–based economy.[79]

The changes required to keep global warming below 1.5°C as outlined by the IPCC are substantial. They require both halting the emissions of GHGs and the adoption of some negative-emissions technologies that would remove the GHGs already in the atmosphere.[80] To date, these technologies have yet to be proven to work at the scale necessary. This degree of social change is impossible without a radical transformation in the ways that the state, market, and civil society sectors interact. The most common examples of drastic social change in response to crisis are war, natural disaster, and economic depression.[81]

Barring a world war or widescale economic depression, both of which have been noted in the IPCC's most recent report as possible consequences of climate change,[82] the type of radical social change needed is most likely to be initiated by civil society. A global mass mobilization that employs either nonviolent or more confrontational tactics has the potential to motivate the type of social transformation needed. Nonviolent conflict has been found to be successful in bringing about such large-scale social transformations if a critical mass of 3.5 percent or more of the population participates in the activism.[83] Beyond responses to repressive and autocratic rule, however, examples of sustained activism at this level of engagement are scant. Thus, it is unrealistic to imagine that such a high percentage of the population would mobilize and engage in peaceful climate activism without some sort of large-scale risk pivot to motivate it.

It is possible that less peaceful forms of activism could lead to the kinds of radical social changes that are needed to address the climate crisis. In fact, calls for such confrontational activism are growing.[84] To date, however, we are nowhere near the level of mass mobilization that is expected to be associated with the types of systemic social change required to respond to the climate crisis. It is also worth noting that there is a continuum of activism that ranges from peaceful to violent protest. Even when activism starts out peaceful, confrontational tactics like roadblocks, sit-ins, and human barriers that are designed to be nonviolent run the risk of turning violent due to interactions with counterprotesters, or the response of law enforcement; repression can easily lead to an escalation of violence on both sides.[85]

I call myself an apocalyptic optimist. In other words, I do believe there is hope to save ourselves from the climate crisis that we have caused. However, I also believe that saving ourselves will only be possible with a mass mobilization that is driven by the pain and suffering of climate shocks around the world. A generalized sense of extreme risk can lead to peaceful and less-peaceful mass mobilizations at the levels needed to stimulate an AnthroShift. Only a global risk event (or numerous smaller events that are seen as threatening social and economic centers of power) will motivate the kind of massive social change that is needed. In other words, without a risk pivot—be it driven by social or environmental change—an AnthroShift that is large enough to respond adequately to the climate crisis and open a large enough window of opportunity post-shock is improbable.

At this point, it is impossible to predict if such a shock will come from ecological disaster, war, pandemic, or another unforeseen risk. What is certain, though, is that without such a shock that motivates an AnthroShift large enough to reorient all the sectors of society to respond meaningfully to the climate crisis, it is hard to envision the world achieving the levels of climate action needed. Instead, the best we can hope for is incremental change that does not disrupt the dominant nodes of political and economic power; such incremental change has the potential to *reduce* the gravity of the crisis, but it *will not stop the coming climate crisis.*

TAKING BABY STEPS TO SAVE OURSELVES

This book is my effort to harness the findings from more than twenty-five years of research to think through how we can save ourselves. It looks at where we are, where we need to go, and what are reasonable paths to get there. Rather than leaning into wishful thinking that a still to-be-developed technological innovation can save us from the climate crisis, I extrapolate from my years of structured research and observations that find that international and national efforts to address climate change are not going far enough.[86] Instead of waiting for a slow and ineffective climate regime to save us, we need to identify and acknowledge our power, then figure out how to harness it effectively so we are prepared to survive what's coming.

As I lay out in detail in this book, there is insurmountable evidence that current responses to the climate crisis from the state, market, and civil society sectors will not save us from a warming world. While we scramble to agree to and implement policies that all interests and actors support, carbon concentrations in the atmosphere continue to rise and the effects of climate change continue to be felt as climate shocks with increasing intensity and growing frequency. Although it is possible that an unrelated risk—such as another pandemic—could motivate social change that puts us on a better path, the COVID-19 pandemic shows us that, even when big social change happens during periods of heightened risk, systemic change requires a sustained sense of risk and threat of shock.

Given our current trajectory, there is ample evidence that we have changed the world and that it will get worse before it gets better: it is likely that only an AnthroShift driven by a substantial risk pivot due to climate shocks and the economic disruption, mass migration, and human suffering that it will cause can put us on a more sustainable path. *Saving Ourselves* provides an overview of what we can realistically expect and considers how far we need to go to get there.

This book walks us through a path forward and focuses on how to make the coming shocks count. In chapter 2, I present an overview of the governmental and business responses to the climate crisis. I provide evidence from years of research on climate policymaking in the United States and abroad to explain why change has been so slow and why the incremental changes we have observed so far are insufficient. Chapter 3 turns to the role that the people must play in bringing about sufficient climate action. It begins with a discussion of the ways that individual action can contribute to saving ourselves and then clarifies why individual action alone is not enough to address the problem. The chapter then discusses how collective action through activism has been working to respond to the climate crisis. I discuss why the outcomes of these efforts, like those of the state and market sectors, have not been adequate to meet the challenge of the climate crisis. Chapter 4 focuses specifically on the growing radical flank of the climate movement, which is using shock and disruption to get attention for the climate crisis and push for more aggressive climate action. I discuss the rift in the climate movement that motivated the growth in the radical flank. Then I explain the different kinds of direct action that are being used to shock and disrupt. Finally, I present some important lessons learned from the civil rights movement to think through how direct action could be leveraged to help us save ourselves.

Chapter 5 synthetizes the findings from the book to look at how climate shocks can lead to climate action by motivating an AnthroShift. Only a massive shock that is both severe *and* durable can open the window of opportunity wide enough that we will achieve the systemic changes that are needed to attain sufficient climate action. Chapter 5 concludes by discussing opportunities to prepare ourselves by creating community, capitalizing on moral shocks, and cultivating resilience. As this concluding chapter explains, there is still time to save ourselves, but it is not going to be easy—and it is not going to be painless.

Chapter Two

SAVING OURSELVES IS A LONG GAME

WHY OUR INSTITUTIONS KEEP FAILING TO ACT ON CLIMATE

ON THE day before the UN Climate Change Conference of the Parties' twenty-seventh round of climate negotiations (COP27) began in the isolated resort town of Sharm el-Sheikh, Egypt, in November 2022, *The Economist* published an article summarizing the sad state of climate policymaking. The headline declared: "The world is going to miss the totemic 1.5°C climate target; it needs to face up to the fact."[1] This article in a publication well known for its centrist political leanings echoed the dire messages coming out of the scientific community that produced the Intergovernmental Panel on Climate Change (IPCC) *Sixth Assessment Report*,[2] as well as the reports from various programs at the United Nations. All were expressing their concern in the leadup to this international meeting. Ten days before COP27 began, in fact, the UN Environment Programme published its annual *Emissions Gap Report*, which concluded that "the international community is falling far short of the Paris goals, with no credible pathway to 1.5°C in place. Only an urgent system-wide transformation can avoid climate disaster."[3]

These messages of failure were even more pronounced once the negotiations got underway. In his opening speech, UN Secretary-General António Guterres declared: "We are in the fight of our lives and we are losing . . . and our planet is fast approaching tipping points that will make climate chaos irreversible . . . We are on a highway to climate hell with our foot on the accelerator."[4] The secretary-general went on to encourage countries to commit to phasing out coal dependence, tax the profits of fossil fuel companies, and redirect those funds to support some of the loss and damage suffered by countries due to climate change during the climate negotiations. This perspective was echoed in the findings from the independent Climate Action Tracker in its report about the additional commitments that were expected in the year leading up to the COP27 meetings but never came: the "process has failed to deliver the urgent emissions cuts governments promised to deliver to keep warming to 1.5°C."[5]

These grim warnings to the international community are in contrast to the messages coming out of the United States. After decades of failed attempts by other presidential administrations, the Biden administration had finally passed a climate bill through both houses of the U.S. Congress in summer 2022.[6] The Inflation Reduction Act (IRA) aimed to achieve numerous goals, including reducing inflation and the deficit, lowering drug prices, as well as addressing the climate crisis by investing substantially in clean energy. Although there are many legitimate criticisms of the IRA—including how it focuses exclusively on carrots rather than sticks, that it does not do enough to protect frontline communities and communities of color, along with the fact that it only passed the Congress thanks to a deal that included the expansion of fossil fuel extraction and infrastructure—there is still much to celebrate about its passage.[7] Most notably, the Inflation Reduction Act is the *only* time in the thirty-plus years since the climate regime began that the U.S. Congress passed any bill through both houses that commits the U.S. government to take steps to address climate change.

Even with this unprecedented political win, the United States is like most other developed countries and still not on track to meet its emissions reduction targets committed to as part of the Paris Agreement in 2015. Although many developed nations, including the United States, have filed Nationally Determined Contributions (NDCs) that indicate plans to reduce their emissions to be more consistent with the IPCC's targets, the implementation of policies that achieve these intended goals are harder to assess.[8] In the 2022 UN *Emissions Gap Report*, national climate pledges going into the COP27 round of negotiations were still "far from the Paris Agreement goal of limiting global warming to well below 2°C . . . Policies currently in place point to a 2.8°C temperature rise by the end of the century."[9]

This chapter tracks the long slog to climate progress to assess why climate action has been so ineffective and why it continues to be insufficient to address the problem.[10] It begins in 1992 at the fabled UN Earth Summit in Rio, which established a series of international environmental agreements that were created to protect the planet. Then it follows the slow and painful journey through many attempts at incremental progress to address the climate crisis over thirty years through the COP27 round of the negotiations in Egypt in 2022. After discussing climate policymaking at the international level, I focus on the primary impediment to climate progress: fossil fuel interests. This section of the chapter assesses the role these interests have played in the slow-walk to any transition away from fossil fuels, including in slowing the clean energy transition by wielding power over policymakers. Finally, I turn my attention to the country responsible for the largest share of historical greenhouse gas emissions: the United States.[11] This section explains how the United States ended up finally passing legislation that addressed climate change at the eleventh hour. It describes in detail how passage was only possible thanks to concessions made to the fossil fuel industry brokered by a senator from the fossil fuel–extracting state of West Virginia.

I end the chapter by looking at what was expected during the 118th session of the Congress, when Republicans took back control of the House of Representatives. Even with groundbreaking investments through the IRA, there is no question that fossil fuel expansion will continue in the United States, and there are reasons to believe it will even increase.[12] Consequently, we should be prepared to experience more frequent and intense climate shocks.

THE UPS AND DOWNS OF INTERNATIONAL CLIMATE POLICYMAKING

There was a time when I would go to all the big climate negotiations; I was even there when the Kyoto Protocol was finished at the second round of the COP6 negotiations in summer 2000. I stayed up all night watching the teams of negotiators finalize the text that made it possible to enter the ratification phase of the protocol. Attending the climate meetings and participating in the side events where groups presented their findings to the world used to make me hopeful. After two weeks of debate, discord, and performative protest, everyone would come together at the last moment with some sort of an agreement that would *finally* work. Attending the meetings felt like we were in the middle of the most important and inspiring race to save the world.

After so many years and too many rounds of last-minute deals to save the world, it's clear to me that the real story of international climate governance is a roller-coaster ride of promising opportunities that ended up falling short of achieving their goals. It has been more than thirty years since the UN Framework Convention on Climate Change was finalized and opened for signatures. Since then, countries have held twenty-seven rounds of negotiations (twenty-eight by the time this book is published) with the goal of implementing policies that will limit the effects of climate change. Over the years, the international regime has cycled through a range of policy instruments—from the completely voluntary to the legally binding. No matter the mechanism, though, all these international agreements have failed to meet their goals of stopping the constant increase in concentrations of greenhouse gases building up in our atmosphere.

Watching all these agreements fail to achieve their broad climate goals in terms of emissions reductions and providing adequate financial support for loss and damage in the developing world has jaded me. After all these years, I have a very different perspective on these huge international meetings that create the appearance of progress without achieving sufficient tangible goals (and emitting too much carbon in the process). They are all just smoke, green mirrors, and a bunch of hot air. No longer do these meetings seem like opportunities to save the world; rather, they are exercises in greenwashing and elaborate games to kick the can down the road until it's too late.

THE EARLY DAYS OF THE CLIMATE REGIME

It all started at the Earth Summit in Rio de Janeiro in 1992 where the United Nations Framework Convention on Climate Change (UNFCCC) was created along with a number of other environmental regimes, including the Convention on Biodiversity and the Commission on Sustainable Development.[13] Likely because the UNFCCC does not include legally binding commitments, the Framework Convention was quickly ratified by nations and entered into legal force (the technical term for becoming an actual treaty) in March 1994. The UNFCCC was the first international treaty on global climate change. The regime recognized that most of the greenhouse gas concentrations in the atmosphere were emitted by industrialization processes in the developed world.[14] As such, the treaty committed developed (so-called Annex 1) countries to stabilize their emissions at 1990 levels by 2000 voluntarily. Developing countries were not bound to emissions reductions in the UNFCCC. By 1997, however, the carbon emissions of developed countries had *risen 9.7 percent above 1990 levels.*

Given the obvious lack of progress, countries decided they needed to create a stronger climate agreement. The Kyoto Protocol was finalized at the third Conference of the Parties (COP3) of the UNFCCC in Kyoto, Japan, in 1997. The protocol was structured to include legally binding (i.e., enforceable) emissions reduction commitments for industrialized countries and those that were classified as economies-in-transition. Like the UNFCCC, developing countries, including the rapidly industrializing China, India, Brazil, and South Africa, were not bound to any emissions reduction targets or timetables. The specific rates of emissions reductions varied by country, but the overall goal of the protocol was to reduce collective emissions by at least 5 percent below 1990 levels by 2008–2012.[15] Although the treaty entered into legal force in 2005, it did so without the support (and ratification) of the country that was the largest greenhouse gas emitter of the Annex 1 nations: the United States.[16]

The "success" of the Kyoto Protocol can be interpreted in a variety of ways. In terms of meeting the emissions reduction commitments of the Annex 1 countries that ratified and stayed in the treaty, it worked well. The UNFCCC secretariat reported in 2014 that the aggregate greenhouse gas emissions compared to 1990s levels "for all Annex I Parties, excluding land use, land-use change and forestry (LULUCF) emissions and removals, decreased by 10.6 percent."[17] However, this information only accounts for the developed countries that stayed in the treaty. During the same time period, U.S. carbon dioxide emissions grew 5.4 percent above 1990 levels.[18] With the United States out of the treaty and not working to meet its emissions reduction targets in the negotiated agreement, other countries were discouraged from following through on their promises. In 2010, Canada, Japan, and Russia removed themselves from any additional commitments, and Canada withdrew from the treaty entirely in 2011, citing "the lack of wider participation."[19] While developed countries were increasingly hesitant to commit to the regime, global emissions climbed substantially. By 2012, the developing non–Annex 1 countries, including China and India, accounted for 59 percent of all global carbon emissions.[20]

THE PARIS AGREEMENT AND BEYOND

As it became increasingly clear that an international climate agreement that only bound some countries would not stop climate change, the international community shifted gears once again and worked to develop a different type of international agreement. In contrast to the legally binding mechanism of the Kyoto Protocol, the next treaty focused instead on individual contributions and was nonpunitive.[21] Eighteen years after the Kyoto Protocol was originally drafted, the Paris Agreement was adopted at the COP21 round of climate negotiations in 2015. The Paris Agreement used a more voluntary structure to engage nations and included a wider range of countries, most notably the top emitters of China and India. This inclusion, along with the fact that the United States had a Democratic president who was committed to the agreement, ensured that the treaty entered into legal force quickly.

The more flexible structure of the Paris Agreement made it possible for the Obama administration to sign on and provide evidence to the Senate that it could meet its commitments through policy implemented via executive order (and not through congressionally approved legislation, which had proven elusive). President Barack Obama introduced the Clean Power Plan as part of his Climate Action Plan in August 2015. The plan was designed to regulate emissions from energy utilities and cut carbon emissions to the level that met the U.S. commitments to the Paris Agreement.[22] With these policies introduced (but not yet implemented), the Senate requirements were met.[23] As a result, the United States ratified the agreement on September 3, 2016, and the international agreement entered into legal force four days before Donald Trump was elected president on November 4, 2016.[24]

The Paris Agreement relies on individual climate commitments—called Nationally Determined Contributions (NDCs)—carried out by individual countries. Every five years, countries are expected to submit updated climate plans.[25] The first updated plans were due in 2020 but were delayed because of the COVID-19 pandemic. Although world leaders were expected to coordinate a global response to the climate crisis that met the challenges the world was facing at the delayed COP26 climate negotiations in Glasgow in November 2021, the outcome was disappointing. Countries were unable to agree to the emissions reductions needed to respond to the dire warnings of Working Group 1 of the IPCC, which released its *Sixth Assessment Report* before the international negotiations began in 2021.[26]

There was ample evidence at the meeting in 2021 that the Paris Agreement was failing: most countries had not yet met their original commitments from the agreement and, even after an extended period of limited travel within and across countries due to the pandemic, global carbon emissions had quickly rebounded. In the words of a report by the International Energy Agency (IEA): "Global energy-related carbon dioxide emissions rose by 6 percent in 2021 to 36.3 billion tonnes, their highest ever level, as the world economy rebounded strongly from the Covid-19 crisis and relied heavily on coal to power that growth."[27] After two weeks of muddled negotiations in Scotland, countries punted and announced that governments should arrive at the COP27 round of negotiations the following year armed with plans for climate action that would limit global warming below the 1.5° threshold that scientists have determined will keep the most dire effects of climate change at bay.

Unfortunately, COP27 in Egypt did not live up to everyone's hopes with regard to mitigating the effects of climate change. After an extended period of negotiating during the final hours of the meeting, the parties were yet again unable to come up with a pathway to limit warming below the 1.5°C target. In addition, despite the explicit priority that the UN chief had outlined in his opening remarks, countries shied away from committing to phase out fossil fuels.[28] In "Climate Change: Five Key Takeaways from COP27," the BBC noted that the final text from the meeting in Egypt did not include language about emissions peaking before 2025, the phase down of coal as an energy source, nor the need to shift away from energy produced by the burning of fossil fuels generally.[29] In short, when we look at climate mitigation, COP27 was a failure.

The silver lining of this round of climate negotiations was the breakthrough regarding how countries experiencing the loss and damage associated with climate change were supported. Given the way the world is experiencing the effects of climate change, developing countries, many of which are in warmer climates and have released a very small proportion of the greenhouse gases that are causing climate change, are feeling the loss and experiencing the damage more quickly and substantially. During COP27, the parties finally followed through on promises to the countries most affected by climate change, ending "decades of rich country refusal to set up a fund to help climate victims in vulnerable countries recover."[30] Even though they committed to create a pooled fund for loss and damage, there is reason to wonder if countries will actually follow through on their pledges to pay into the fund given previous behaviors and the lack of follow through on previous pledges.[31]

While carbon emissions continue on their upward trajectory and governments spin their wheels trying to agree to ineffective international policies, researchers have documented the ways that transnational climate governance could work through polycentric approaches to address the climate crisis.[32] But even with numerous efforts to address the problem at multiple levels of governance, policymaking has not brought about the emissions reductions necessary to limit global warming below 1.5°C. Not only has Earth already warmed more than 1°C,[33] but it is well documented that additional warming is already baked into the system through investments in fossil fuel infrastructure expansion and the time it will take for a transition to clean energy.[34]

After the publication of the Working Group 1 report in 2021, *Nature Magazine* reported the results of a survey of scientists involved in the IPCC's most recent assessment, which found that most of them agreed that 1.5° was beyond our reach.[35] Since 2021, numerous reports have echoed the message that the 1.5° target will be missed.[36] In my meetings with scientists working on climate change inside and outside the U.S. government, I find that most scientists respond to questions about the viability of the 1.5° target with an uncomfortable cough and, in many cases, an eye roll.

For those of us who have been tracking the various rounds of climate negotiations over the years, the results of this most recent round of negotiations were disappointing (as is the growing consensus within the scientific community that the target will be missed). These results are also not particularly surprising. At the same time, the meetings of the climate regime have become less-and-less transparent and inclusive of civil society actors.[37] This trend continued at the COP27 meeting, which was held in the authoritarian state of Egypt and severely limited participation by nongovernmental organizations (NGOs).[38]

While participation by civil society actors was limited, the COP27 meeting boasted the highest level of fossil fuel representation ever at a climate meeting.[39] With such a trend, it should not be a surprise to find the parties at the negotiations unable to commit to the phaseout of fossil fuels. It is hard to imagine that this trend will change when the COP28 round of negotiations takes place in the fossil fuel state of Dubai in fall 2023. Perhaps Damian Carrington, the environment editor of *The Guardian*, said it best in his postmortem of the Egypt round of the negotiations in 2022: "When the history of the climate crisis is written in whatever world awaits us, COP27 will be seen as the moment when the dream of keeping global heating below 1.5C died."[40]

As emissions climb, temperatures continue to rise, and nations continue to hold meetings in far-off locales to discuss how to implement an international treaty that might achieve the necessary emissions reduction goals. Figure 2.1 provides a recap of the effects of thirty years of negotiating for climate action. It overlays the timing of some of the biggest rounds of the international climate meetings on top of the concentrations in atmospheric carbon dioxide and global temperature change over time. The figure uses the "climate stripes" developed by Ed Hawkins at the University of Reading to depict the rise in average temperature.[41] It provides a clear message about where we are globally in terms of two of the most common indicators of climate change: temperature and carbon concentrations in the atmosphere. In short, it shows we are nowhere near where we need to be.

UNPACKING THE INFLUENCE OF FOSSIL FUEL INTERESTS

Even though the climate regime and the UN-sponsored international scientific assessment have avoided explicitly discussing the role that fossil fuels have played our changing climate, it is impossible to tell a comprehensive story about climate policymaking without talking about the powerful economic interests

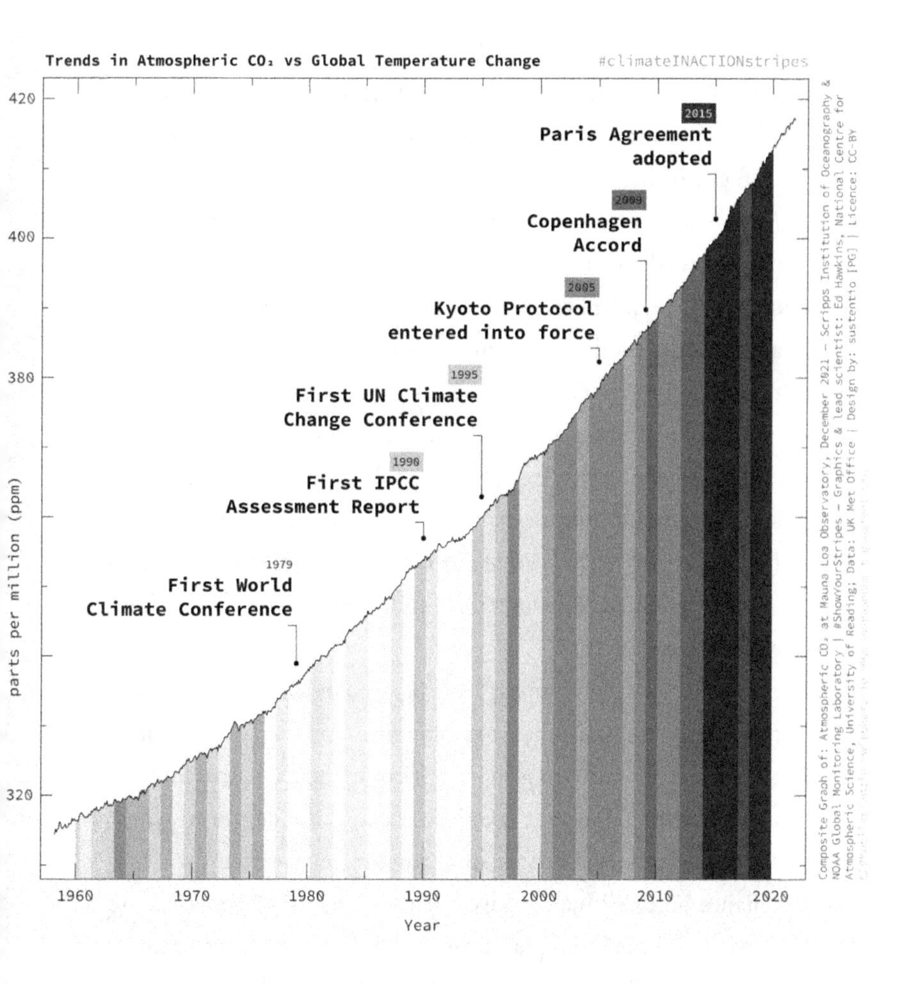

FIGURE 2.1 Atmospheric CO2 concentrations, global temperature change, and rounds of climate negotiations in one figure. sustentio, "Feel Sustainability," https://sustentio.com/2022/climateinactionstripes-virale-klimakommunikation (accessed June 26, 2023).

that profit from the extraction and combustion of fossil fuels. These interests—which go beyond the companies that extract coal, oil, and natural gas—are well documented for blocking any shift away from the system that is fueled by burning fossil fuels. In her contribution to Greta Thunberg's *The Climate Book*, Naomi Oreskes provides a succinct overview of how this process works in reality by summarizing years of her research on the subject. In her own words, the lack of sufficient climate action is the product of a "history of denial and obfuscation by the fossil fuel industry . . . The fossil fuel industry and its allies acted indirectly to prevent climate action by poisoning the well of public debate, but they also acted directly when government action appeared imminent."[42] Fossil fuel interests have played a role in slowing down climate progress at the international level as well as within nation states (and regions) where they hold concentrated power.[43]

While scholars coming from an ecological modernization perspective suggest that technological innovation will drive the necessary shifts to a more environmentally sound future,[44] this viewpoint does not take into account the degree to which some interests—like those that support and maintain society's reliance on fossil fuels—have been granted privileged access to power since the industrial revolution.[45] These entrenched interests will use all the tools at their disposal to slow down an energy transition and to maintain power. Thus, business responses to climate change have not universally shifted as clean energy has become cheaper and more available.

As I noted in chapter 1, business responses instead have been bipolar. On the one hand, as the clean energy sector grows and the prices go down, more and more companies are supporting policies that encourage the transition to clean energy. At the same time, entrenched fossil fuel interests continue their efforts to slow the energy transition at all costs. To date, only efforts to implement policies that support fossil fuel interests (by either expanding their capacity to extract and burn more resources or providing opportunities for them to profit from carbon removal) have garnered broad support in the United States.

Due to the push-and-pull of these different market actors, social responses to the climate crisis do not follow the same trajectory as the swift and effective global response to ozone depletion. In contrast to the solution to the ozone crisis, which was unique in many ways, companies representing clean (i.e., non-carbon-emitting) energy sources and technologies continue to compete with entrenched business interests that support an economy fueled by fossil fuels. A growing literature documents how these vested economic and political interests have worked to maintain the status quo through the process of agency and regulatory capture,[46] as well as by employing dark money that funds and disseminates climate misinformation that aims to confuse and distract from climate action.[47] In 2023, for example, numerous so-called citizen groups expressed concerns about how the construction of wind power off the coast of New Jersey might threaten whales. However, they were exposed as funded by oil industry interests and connected to a right-wing think tank that was focused on shifting public opinion away from renewable energy.[48]

In other words, these concentrated interests continue to exercise a disproportionate amount of power over decision makers.

They have been credited with successfully muting the message of the various IPCC reports, including the most recently released *Sixth Assessment Report*, which downplayed the role of fossil fuel companies in the climate crisis in the text of the report, as well as in the Summary for Policymakers.[49] Figure 2.2 from the Global Carbon Project shows the direct relationship between overall carbon emissions and those coming from burning fossil fuels. It is pretty clear from the figure why those vested interests have resisted climate action for so long: any meaningful response to the climate crisis must involve substantially reducing our reliance on fossil fuels. The only alternative option is that a technology that removes carbon, either at the utility or directly from the atmosphere, becomes cheaply available and capable of working at scale. To date, however, no such technology can work at this level (but billions of dollars are being spent to try to make this dream a reality).[50]

It is important to keep figure 2.2 in mind as we discuss climate inaction at various levels of governance. Given the disproportionate power that fossil fuel interests hold over capital, it should not be a surprise that a November 2022 article in *Energy Monitor* notes how global "clean energy investment is only marginally increasing, while the financial system is locked into autopilot with hundreds of billions in new investments in fossil fuel infrastructure every year."[51] The "2023 Statistical Review of World Energy" found that, even with renewables going up, fossil fuel consumption remained steady at 82 percent of primary energy.[52] In the United States, natural gas production has also been growing.[53] Even with the IRA and the Bipartisan Infrastructure Law, the Biden administration has also been

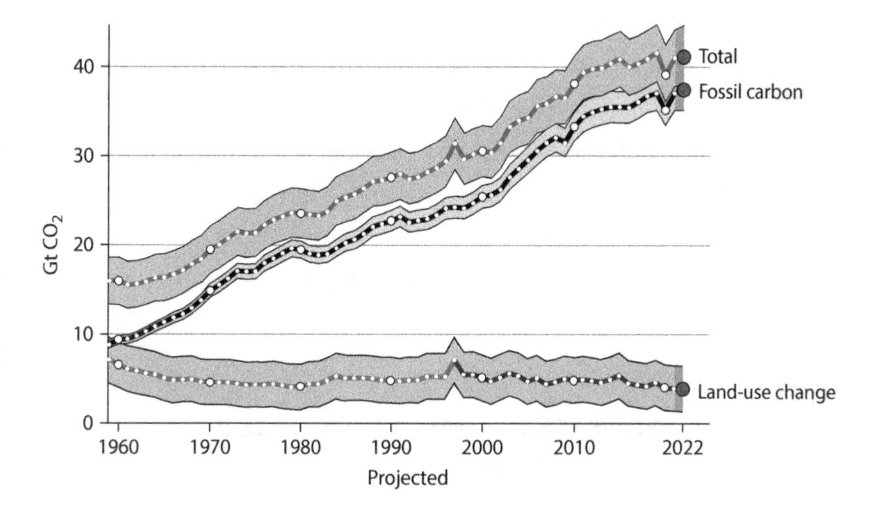

FIGURE 2.2 Annual carbon dioxide emissions. Robbie Andrew, "Figures from the Global Carbon Budget 2022," from Pierre Friedlingstein, et al., "Global Carbon Budget 2022." *Earth System Science Data* 14, no. 11 (2022): 4811–4900. doi: 10.5194/essd-14-4811-2022.

supporting the expansion of fossil fuels.[54] Moreover, although many fossil fuel companies released what they call "zero emissions" plans to reduce their emissions, these plans have been criticized for not including the emissions that come from the *burning* of the fuels that are sold to other companies and/or countries.[55] As a response to such plans, the UN issued a high-level report that focuses on the loopholes in these types of plans, which it calls "greenwashing." The report notes, "actors cannot claim to be 'net zero' while continuing to build or invest in new fossil fuel supply or any kind of environmentally destructive activities. They can't also participate or have their partners participate in lobbying activities against climate change or just report on one part of their business's assets while hiding the rest."[56] After reporting huge profits, many oil and gas companies announced more recently that they would delay their plans to phase out fossil fuels (which had already been criticized for being too slow).[57] Given their watered-down pledges and willingness to delay their climate goals, these efforts by businesses should be interpreted as predominantly smoke and green mirrors.

PERSISTENT CLIMATE BLOCKAGE IN THE UNITED STATES

We can see clearly how fossil fuel interests have slowed progress toward climate action by looking at the country that has contributed the most carbon to the atmosphere over time: the United States. The country, which has expansive stores of coal along with petroleum and natural gas[58] "is responsible for the largest share of historical emissions . . . with some 20 percent of the global total."[59] During his opening remarks at a climate workshop at the Harvard Kennedy School in December 2022, Joseph Aldy, who served as special assistant to the president for energy and environment during the Obama administration, reviewed the history of failed efforts to regulate carbon emissions. In his own words, U.S. history is littered with the "roadkill" of numerous failed climate policies.[60]

Over the thirty years since the Earth Summit in 1992, the United States has shown itself to be particularly slow in addressing climate change. As the political party in power has changed, its formal stance on the issue has flip-flopped more than once. Not only did President George W. Bush deviate from the Clinton administration's position and decide not to move forward on the Kyoto Protocol in 2001,[61] but the United States also formally withdrew from the Paris Agreement, which President Obama had originally negotiated, when the Trump administration took office. During the first week of the Biden administration in 2021, the United States formally recommitted to the international treaty. This variation in national position on the issue is not unique to the United States; other resource-rich developed countries—including Canada and Australia—have faced similar challenges.

When we dig deeper than the country's formal position on the climate issue, we can see the many ways that geographically diffuse fossil fuel interests have swayed political outcomes in the United States. The connections between the fossil fuel sector and elected officials have been found to play a substantial role in their political positions. Elected officials who represent areas rich in fossil fuel resources tend to vote based on the interests of the resources in the areas they represent.[62] Perhaps the most well-known example is Senator Joe Manchin, who holds the seat formerly held by Robert Byrd from the coal state of West Virginia. Byrd was renowned for regularly pushing to support funding for "clean coal" technology as part of the federal appropriations process. He was also famous for his role in personally guaranteeing the success of the Byrd-Hagel Resolution, which blocked Senate support for the international climate agreement that became the Kyoto Protocol in 1997.[63]

Although members of the Republican Party are well known for taking funds from coal, oil, and natural gas companies and voting against climate action, fossil fuel interests have long supported political campaigns *on both sides of the aisle.* They also provide campaign funding for officials who come from areas without large natural resource endowments. There is clear evidence that elected officials who receive this type of funding—no matter their party affiliation or whether they represent areas that extract fossil fuels—are more likely to vote against environmental policies, including those aiming to address climate change.[64]

My previous research documents the failure of numerous national efforts to address the climate crisis. I have studied legislative efforts, including the Climate Stewardship Act of 2003,[65] which did not pass the Senate, and the American Clean Energy and Security Act of 2009, also known as the Waxman-Markey bill,[66] which passed the House but failed in the Senate in the early days of the Obama administration. I have also researched administrative efforts like the Obama administration's Clean Power Plan, which was withdrawn after the Trump administration took office.[67] Across all these policies, the fingerprints of fossil fuel interests can be seen if you look hard enough. Given the role that these interests have played in blocking federal climate action over the years, there is reason to expect that the only way a bill intended to address climate change could pass through the U.S. Congress is if it also provided incentives to traditional fossil fuel–based energy interests.

TO BUILD BACK BETTER—OR NOT

The Biden administration came into office with the promise of taking aggressive climate action through what it called the

Build Back Better plan. Joseph Biden the candidate had run on a platform that included the most comprehensive climate plan ever, which included a pledge to "end fossil fuels."[68] Although young people did not initially support candidate Biden, my research on the youth climate movement during this period of the 2020 election documented how they lined up to support him once he became the nominee.[69] After winning the nomination, Biden's campaign worked with various climate groups to finalize his platform.[70] These groups included the youth-led Sunrise Movement that had gained notoriety when it occupied Nancy Pelosi's office in 2018. In exchange for numerous additions to the Biden campaign's climate plan, climate groups provided campaign support. For example, the Sunrise Movement coordinated youth-led phone- and text banking efforts for candidate Biden in the swing states of Arizona, Florida, Michigan, North Carolina, Pennsylvania, Texas, and Wisconsin during the election.[71]

The Build Back Better Plan was born out of President Biden's climate agenda, along with policymaking insights garnered from decades of failed attempts to pass climate legislation in the United States. In contrast to previous efforts like the Climate Stewardship Act of 2003, the American Clean Energy and Security Act of 2009, or the Clean Power Plan, all of which aimed to address climate change by limiting carbon emissions of the industrial sector and/or power plants with regulations and market-based efforts, this Plan addresses the climate crisis by investing in the clean energy sector in the United States. While it does not put a price on carbon or create financial incentives to limit carbon emissions, it provides numerous economic incentives to shift to clean energy sources. In other words, it is a climate policy that is based on carrots, not sticks.

The Plan took advantage of Senate rules around the budget reconciliation process that means "it cannot be filibustered, and only needs a simple majority to pass."[72] Since the Democratic Party had only a one-vote majority in the Senate at the time, the reconciliation package had to be agreeable to all Democrats to make passage possible. As a result, certain senators were in a privileged position to affect the content of the bill. Most notably, Democratic Senator Manchin from West Virginia was able to hold the process hostage for over a year until he got what he wanted.

After Manchin publicly rejected the Senate's version of the House-passed Build Back Better Act on Fox News in December 2021, it seemed like dreams of the climate-focused legislation were dead in the water and all that time spent negotiating to get a Senate companion bill that was passable was a complete waste of time. It was so acrimonious, in fact, that when I hosted an event at the Library of Congress in June 2022 to talk about clean energy transitions in the United States, a congressional staffer referenced the universally reviled villain from the Harry Potter books, calling it "the bill that will not be named." The comment was met with uncomfortable laughter by the politicos around the table.

Despite this rancor, a less generous bill was resurrected later in summer 2022. Manchin and Senate majority leader Chuck Schumer surprised many when they announced a deal for a trimmed-down reconciliation package—called the Inflation Reduction Act—in August 2022, three months before the midterm elections. The bill directly connected renewable energy development to fossil fuel expansion: the government must offer federal land for fossil fuel leasing as a precondition to wind and solar projects being leased and developed.[73] Entangling fossil fuel expansion to the climate provisions in the bill led journalist Kate Aronoff, in an article for *The New Republic*, to call the IRA "both a big win and a deal with the devil" that "could be a boon both for fossil fuel executives' core business model and for their prospects of profiting from the clean energy technologies that might eventually replace them."[74]

To get Manchin to support the package, Schumer also agreed to help Manchin gain approval for the Mountain Valley Pipeline that would transport natural gas from West Virginia. Although it took ten months, the provisions were included as part of the debt ceiling bill that was signed by the president in June 2023.[75] Between this side agreement and the many giveaways to fossil fuel interests included in the bill, there is a lot of evidence that the IRA passed through the Senate (and did not become more roadkill) only because it was bundled with the expansion of fossil fuel infrastructure.

Finally, after more than twenty years of federal efforts to pass legislation that addresses climate change punctuated by periods of resistance to climate policymaking when the Republican Party held power, a bill that invested heavily in addressing the climate crisis passed the Senate on August 12, 2022. President Biden signed the IRA flanked by members of Congress, including Manchin, later that month. Even with its many weaknesses, the passage of this bill means that the United States finally has a federal climate policy that aims to reduce carbon emissions.[76]

The fact that the bill passed through the Congress and is legislation that was signed by the president is important because it means that a new president cannot come into office and cancel the policy (as happened with President Obama's Climate Action Plan that included the Clean Power Plan). When I spoke with Jason Walsh, executive director at the BlueGreen Alliance, a coalition of labor unions and environmental organizations, he explained other benefits of legislation over regulations implemented through executive order. "You [can] do things with legislation that you can't do via regulation: you can target investments, invest in the right things, condition those investments with labor standards, etc., in legislation."[77]

WHAT IS THE EFFECT OF THE IRA?

Since its passage in summer 2022, billions of dollars have been distributed to governmental agencies, businesses, and citizens to encourage a clean energy transition, as well as to companies trying to scale up carbon capture and removal technologies. It is unclear, however, if the IRA will lead to the emissions reductions committed to by the Biden administration in its updated NDC to the Paris Agreement. After the bill's passage, the Climate Action Tracker changed the classification of U.S. domestic climate policy from insufficient to "on the border between 'Insufficient' and 'Almost sufficient.'"[78] Without additional climate policies, the most common estimates put emissions higher than the U.S. target of 50 to 52 percent below 2005 levels.[79] While some scholars have interpreted the passage of the IRA as "the beginning of a green transition" in the United States,[80] others are skeptical that the policy will motivate the rapid transition away from fossil fuels that is needed, especially as the Biden administration approved more fossil fuel projects in its first two years in office than did the Trump administration.[81] The unfortunate reality is that only after it is fully implemented will we be able to assess how effective the law is at reducing carbon emissions and meeting the Paris Agreement commitments.[82]

With Republicans taking back leadership in the House of Representatives in 2023, the momentum on climate policymaking at the federal level slowed; there is no question that any proposed climate policy will face much larger hurdles to passage in the 118th Congress. In figure 2.3, I present findings from my research on the political elites who work specifically on federal climate policy in the United States, or what we call the climate policy network, which was conducted in spring 2022. The figure maps out responses to questions about climate-related debates, proposals, and decisions in the United States. Respondents were asked to indicate their organization and/or office's level of support from strongly disagree (1) to strongly agree (5) to a series of salient policy issues. In the figure, responses to these questions are plotted by policy actor (the black diamonds represent the overall mean for each response).[83]

When we map political elites' perspectives regarding a transition away from fossil fuels by the various policy actor types, the level of disagreement around this topic reveals itself (this statement is circled in the figure). Not only do the Republicans in Congress (the gray circles) oppose a transition away from fossil fuels, Democrats in Congress and the Biden administration (the gray squares) are on the opposite end of the distribution.

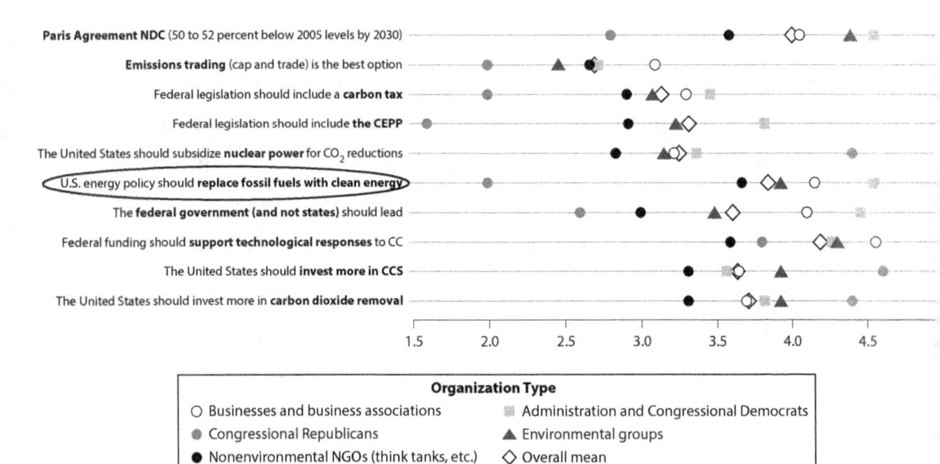

FIGURE 2.3 Climate-related policy stances by actor type. Dana R. Fisher, "What to Expect on Climate Change from the New Congress," Brookings Institution, 2023, https://www.brookings.edu/blog/fixgov/2023/01/03/what-to-expect-on-climate-change-from-the-new-congress/ (accessed February 4, 2023).

In fact, the responses to "U.S. energy policy should replace fossil fuels with clean energy" had the largest standard deviation of all the policy questions we asked.

This difference in perspective was also highlighted during my conversations with U.S. policy elites when I interviewed them in spring 2022. While Democrats and members of the Biden administration discussed how best to transition the United States away from fossil fuels toward more renewable energy sources, Republicans had a different perspective. When I sat down with representatives in the key congressional committees that work on climate and energy, Republican members of Congress told me that expanding American natural gas extraction and exporting it abroad was the best way to address the climate crisis. A staff member for the Republicans on one of the House committees that works directly on energy best summarized this perspective: "The single biggest thing that we can do in the near-term from a climate standpoint is to replace our cleaner, natural gas . . . and supplant much dirtier Russian natural gas."[84]

With a Republican-led House of Representatives, we should expect the new leadership to focus their climate efforts unambiguously on fossil fuel expansion to support increased natural gas extraction in the United States. We can see this priority play out when we look at conversations in the 118th Congress. At the first meeting of the House Energy and Commerce Committee, for example, the Republican chair Cathy McMorris Rodgers from West Virginia, stressed this point: "We need to be doing more to secure and unleash American energy."[85] The conversation focused on simplifying the permitting process and potentially rewriting the National Environmental Protection Act (NEPA), which many Republicans said was responsible for limiting the growth of the energy industry and infrastructure. Similar arguments were made when the Biden administration was considering whether to approve the Willow Project, which would open a large swath of land in Alaska to oil extraction. The *Washington Post* quoted Alaska senator Dan Sullivan, who publicly questioned how President Biden could justify blocking the Willow Project when it had lifted sanctions to allow oil imports from Venezuela, which he called "one of the most polluting places to produce oil anywhere in the world."[86]

Given the expansion of fossil fuel projects in the United States, it is hard to imagine that the United States will meet its commitments under the Paris Agreement by 2030 even with the IRA. Thus, we can expect more warming to occur, which will lead everyone to experience climate shocks that are more severe and more frequent. Keeping in mind this sobering reality about how institutions are failing to meet the challenge of the climate crisis at the level that is needed, I focus on civil society in the next chapter. In it, I turn my attention to the role that individual citizens and civic groups are playing in response to the climate crisis and the ways they might help us save ourselves.

Chapter Three

SAVING OURSELVES INVOLVES TAKING POWER BACK FOR THE PEOPLE

2023 STARTED with a bang: after the ball dropped on an unseasonably warm New Year's Eve in New York City, abnormal weather was experienced around the globe. In January, the unusually warm winter closed many ski resorts on the continent,[1] while the Western United States experienced a series of climate change–amplified atmospheric rivers that triggered mudslides and flooding, along with historic snowfall in the mountains across California.[2] By February, the sea ice around Antarctica had reached the lowest level ever recorded[3] and New Zealand declared a state of emergency after being slammed by multiple cyclones that were intensified by warmer waters.[4]

In the Washington, DC, area, we were also feeling the effects of a changing climate. By mid-February, the University of Maryland had called the winter of 2023 the *year without a winter*,[5] while my hyacinths were blooming and the fabled cherry trees in the District of Columbia were starting to pop with color—over one month earlier than usual! Although some people enjoyed the shorter winter with no ice storms, I found it difficult not to think about how weather that was thirty degrees above average would feel come summer. Lest we thought it was a fluke, extreme weather continued through spring and summer of 2023, with the Pacific Northwest of the United States and Western Canada reporting a heat wave in May that broke records by over 10°F in some places and an unprecedented wildfire season in Canada that blanketed many parts of the United States in hazardous smoke.[6]

While we continue to grapple with the effects of a warming world, and climate shocks grow more severe and come more frequently,[7] policies are moving too slowly to meet the crisis. As a result, civil society—the sector of society that is outside the state and the market and includes all of us, everyday people, acting based on our personal identities, interests, and concerns[8]—has become increasingly focused on climate change. Worry about the issue varies by country, but numerous sources have documented how concern has grown.[9] Like other countries, public concern for climate change continues to increase in the United States. Also like people in other countries, the perspectives of Americans vary substantially based on the individuals' political orientation and age—with the youngest generations the most concerned.[10]

In the United States, those who identify as politically conservative remain the least worried about the issue, even though they tend to live in areas most affected by climate shocks.[11] Conservative communities are not only experiencing the effects of the changing climate firsthand, but they are also disproportionately benefitting from federal climate policies including the Inflation Reduction Act (IRA), which is budgeting more than $20 billion to build green energy infrastructure across the United States.[12] As the climate crisis worsens and the economic benefits of public investments in clean energy spread, there is reason to expect that people's perspectives will also change. It is unclear, however, how quickly these changes will come. And even when they do come, decades of research has documented that the road from concern about an issue to civic action can be quite long.[13] Scientists from a broad range of disciplines have stressed that we are running out of time.[14] The window to take action to keep global temperatures below the 1.5° warming threshold is quickly closing.[15]

This chapter focuses on how "the people" have taken power to prevent the worst of the climate crisis. It looks specifically at how individuals and groups have worked to change their personal behaviors as well as to pressure businesses and governments to shift their policies and adjust their practices. I synthesize research from over twenty-five years studying the climate movement to understand how it has grown and the ways it aims to exert social and political pressure. This chapter also builds on the work I did as a contributing author in the most recent Intergovernmental Panel on Climate Change (IPCC) report on mitigation (IPCC WG3), where I contributed to the section about activism and engagement on national and subnational policies and institutions.[16] I unpack how individual and collective action works both directly and indirectly through civic groups to target nodes of political and economic power. Building on my research studying activism and civic engagement, I present a profile of the people who are participating in climate activism in terms of who they are, what they are doing, and what is motivating them to stop the climate crisis.

TAKING POWER AS INDIVIDUALS (AND WHY INDIVIDUAL CLIMATE ACTION CAN'T SAVE US)

Ever since President Jimmy Carter put on a cardigan and told everyone to turn down their individual thermostats,[17] many have leaned into personal behaviors to reduce energy consumption. This call for individual action by the president in February 1977 was motivated by the oil crisis of the 1970s. Attention has refocused more recently on the role that individual behaviors play in energy consumption as part of the necessary responses to climate change.

There is no question that individual behavior change is an important tool in the arsenal to stop the climate crisis. Scholars have stressed the potential of personal behavior change for many years.[18] In their 2008 article, for example, Gerald T. Gardner and Paul C. Stern provide an overview of the research and present a short list of effective actions that individuals can take within their homes to reduce energy as well as carbon emissions. Kimberly Nicholas and colleagues have updated and expanded this work on individual behavior change. Although they explicitly recognize the role of governments in climate policymaking, the authors stress that individual actions can help to mitigate climate change.[19] The authors list four high-impact behaviors that will substantially reduce individual carbon emissions:

having one less child, living without a car, avoiding air travel, and eating a plant-based diet.

While this research focuses specifically on individual behaviors, it also recognizes that the effects of these individual actions vary by country because of the variations in political and economic systems. In Norway, for example, almost all electricity is generated by renewable hydropower (the International Energy Agency [IEA] reports it was 92 percent in 2022).[20] Because their electricity sector does not emit substantial amounts of carbon, the Norwegian government has focused its efforts on reducing carbon emissions from its transportation sector, which accounts for almost a third of the country's emissions.[21] The government implemented a tax system to incentivize consumers to switch to electric vehicles (EVs): cars that are powered by oil are taxed at a higher rate so that the price of electric cars is very competitive.[22] This policy has been quite successful: 79 percent of new passenger cars sold in Norway in 2022 were EVs.[23] This shift to EVs charged by carbon-free hydropower (in contrast to cars with combustion engines that run on petroleum) has led to substantial reductions in individual emissions in Norway.

Although this transition away from fossil fuel–powered cars is great news for Norway and its climate goals, such incentives only work in countries that generate most of their electricity from clean, non-carbon-emitting sources. The United States, in contrast, generates the majority of its electricity by burning fossil fuels (60.2 percent of electricity in 2022).[24] As a result, most EVs in the United States, which accounted for only 6 percent of the market share in 2022,[25] are charged by electric systems powered by natural gas and coal. Not unlike most other industrialized countries, the United States depends on a system that emits carbon to provide energy for businesses and homes as well as to power its transportation sector. Without a transition to non-carbon-emitting energy sources and transportation options that include expanded opportunities for public transit, a transition to electric cars will not be enough, by itself, to stop climate change.[26] It is important to recognize that transitioning to an EV that gets its energy from a grid powered by fossil fuels is a valuable first step in a clean energy transition.

As the variation in EV use in the United States and Norway highlights, individual-level actions can have visible and measurable effects on personal emissions, but they are mediated by political and economic systems that determine the degree to which they help combat the climate crisis. Unfortunately, Norway is one of only a small number of countries that does not rely on fossil fuels for most of its electricity. Mitigating the most harmful effects of climate change in most countries involves changing the way that energy is generated and used to fuel all aspects of life as we know it.

Even in a country like Norway that is mostly powered by clean energy, shifting individual consumption patterns can be quite difficult. In a series of papers based on data from a nationally representative survey of the Norwegian population, my colleagues and I document the limited role that climate considerations have played in choices regarding car use, leisure air travel, and red meat consumption.[27] Our findings highlight two important facts. First, without government intervention, it is very hard to change individual behaviors. Second, although individual behavioral change can be an important tool in efforts to reduce carbon emissions, changes to the political and economic systems are needed to stop the climate crisis. Anyone who says that individual consumption choices can do it alone is trying to distract from where the power is actually concentrated.

This distraction by powerful interests has gotten a lot of attention in recent years.[28] Not only did fossil fuel companies hide their knowledge about the effects that the growing concentrations of carbon in the atmosphere would have on the climate, they invested in efforts to redirect the focus of climate action away from themselves to the individual consumer.[29] As Mark Kaufman reported in 2020, fossil fuel companies developed and promoted the notion of the *carbon footprint* as a means of redirecting civic efforts away from the systemic level. Writing about what he calls the "devious fossil fuel propaganda we all use," he explains how the notion of the carbon footprint helps individuals think about what they can personally "do about the climate problem. No ordinary person can slash 1 billion tons of carbon dioxide emissions. But we can toss a plastic bottle into a recycling bin, carpool to work, or eat fewer cheeseburgers."[30] Investigative journalist Amy Westervelt expanded on this point in a piece in *Rolling Stone:* "Big Oil is trying to make climate your problem to solve. Don't let them."

The inherent fallacy of focusing our efforts exclusively on changing our individual behaviors is that much of the carbon consumption of an industrialized society is inextricably locked into the system. Unless we go completely off the grid and generate our own power, grow our own food, and do not connect with the rest of society, most of our carbon consumption is the product of the broader systems in which our personal and professional lives are embedded. An engineering class at the Massachusetts Institute of Technology (MIT) provided a case in point when they set out to calculate the carbon consumption of a homeless person in 2007. In their analysis, the class documented how even a destitute American who owns neither a house nor a car and lives in a homeless shelter emitted more than twice as much greenhouse gases than the global average.[31]

GETTING TO SYSTEMIC CHANGE THROUGH COLLECTIVE ACTION

With the knowledge that individual efforts alone are not enough to stop the climate crisis, more and more attention has focused on the need for large-scale systemic change. In the IPCC's 2018 special report on the impacts of global warming of 1.5°C, it included a chapter that explicitly looked at the enabling conditions for strengthening and implementing systemic changes.[32] Unfortunately, the discussion of how to achieve this level of change was not picked up broadly.

By 2022, the recognition that systemic change is needed had become quite common. Right before the twenty-seventh round of the climate negotiations—the Conference of the Parties 27 (COP27)—in November, the UN Environment Programme published its 2022 *Emissions Gap Report*, which announced: "Only an urgent system-wide transformation can avoid climate disaster."[33] As I explained in the last chapter, even with these dire warnings, political and economic actors were still reluctant to follow through on calls for climate action. In her recent summary for Greta Thunberg's book, historian of science Naomi Oreskes explains why. She notes that the system was built and sustained by "people in positions of power and privilege [who] refused to acknowledge that climate change was a manifestation of a broken economic system."[34] Recognizing that this system granted fossil fuel interests privileged access to resources and power in the first place, Oreskes publicly asks why we would expect them to be active participants (or even leaders) in the response to climate change.[35] An effective response to the climate crisis requires an energy transition that would lead fossil fuel interests to lose their privileged position of power.

Fossil fuel interests are deeply embedded in the political and economic system. Rather than working to facilitate system-wide transformations to carbon-free power generation, transportation, and industrial sectors, they maintain their power by distracting consumers with their calls for personal actions that focus on individual carbon footprints while they take advantage of unprecedented tax subsidies that ensure their power will continue for decades.[36]

As it becomes clearer that political and economic systems must change to address the worsening climate crisis, and that governments and markets will not take sufficient action independently, the climate movement has grown. The climate issue is so broad and climate action involves changing the ways that society works, plays, and lives; thus, climate activism is similarly broad and diverse. Activists have mobilized to protect the climate from greenhouse gas pollution generally as well as explicitly fighting to stop the funding, siting, building, and expansion of fossil fuel infrastructure, and mobilizing to enforce the laws that are meant to limit the emissions of greenhouse gases. Some activists have focused their efforts on gaining attention for the ways that fossil fuel expansion causes substantial environmental damage to local communities while simultaneously increasing fossil fuel production and consumption.

Collective efforts have historically been mediated by a range of organizational forms that create or remove obstacles to participation.[37] Civic groups that range from those specifically focused on the environment to those that integrate environmental stewardship into their broader missions have organized their members to work to limit climate change. In recent years, digital technologies—including various social media platforms—have facilitated the spread of activism while they connect countries and cultures.[38]

Civic efforts to stop climate change can have a direct or indirect effect on the concentrations of greenhouse gases in the atmosphere (see figure 3.1). (In the best cases, they have both direct and indirect effects.) Efforts that aim to have *direct* effects on the climate involve activists changing their individual behaviors. Groups coordinate their members to make behavioral and lifestyle changes to reduce their personal carbon consumption directly. These steps involve encouraging members to change their purchasing and consumer behaviors, including reducing car use and air travel, shifting to non-fossil-fuel-based sources of electricity, and eating less dairy or meat.[39] Because the actions are mediated through personal and organizational networks that disseminate information and provide social support through connections with friends, neighbors, colleagues, other students, or coworkers, they are likely to be more successful and have more lasting effects than individual efforts alone.[40]

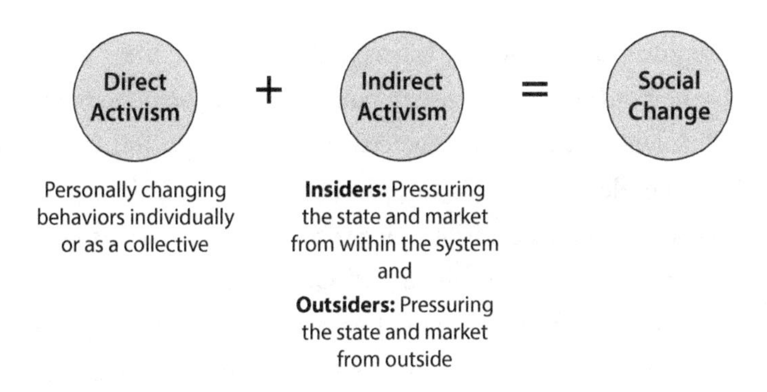

FIGURE 3.1 How civic efforts can contribute to social change

Climate activism can also involve *indirect* efforts to bring about change by pressuring political and economic actors to take climate action. More and more civic groups recognize that changing individual behaviors without changing the system can only go so far. Collective climate efforts are increasingly focused on targeting the behaviors of governments and businesses through a long and *indirect* campaign to achieve climate action.

Activists target nodes of power, including policymakers, regulators, and businesses, with the hope that they can motivate the political will needed to pressure political and economic actors to change their behaviors and/or accelerate their efforts to reduce greenhouse gas emissions. Political and economic behavior change involves enacting the kinds of emissions-reducing policies recommended by scientists working with the IPCC.[41] Building on the legacy of numerous social movements in the United States, activists work both inside and outside institutional systems to apply pressure to nodes of political and economic power.[42]

Insider tactics include activism that aims to affect social change from within the existing political and economic systems. It involves leveraging the judicial system through litigation,[43] lobbying elected officials to influence the policymaking process, and employing various tactics to influence businesses as consumers or even shareholders.[44] Climate activists have also joined recent efforts on the political left to work through the electoral process to change representation through participation in democratic elections.[45] In recent elections, climate groups worked throughout the cycle to identify, endorse, and support climate-sympathetic candidates for political office. With growing evidence that elected officials' voting behavior on environmental issues is strongly associated with whether they receive political funding from fossil fuel interests—no matter the party affiliation of the official[46]—some groups have worked to get climate candidates to sign a no-fossil-fuel-money pledge.[47] After identifying the candidate, helping them develop their climate platform, and providing support for their campaigns, groups then work within the political system to lobby and hold the elected officials accountable once they take office.

The 2020 election of Joseph Biden as president of the United States provides an example of the ways that climate activists work within the political system to encourage climate action. As I mentioned in Chapter 2, although Biden was not initially considered the climate candidate in the 2020 election, his campaign worked with climate groups after he became the Democratic nominee to gain their support. One such group was the youth-led Sunrise Movement, which worked with the candidate to adopt a progressive climate platform that would help secure the support of climate groups and the individual voters who are their members.[48] As a result of these efforts, youth climate groups not only supported Biden's candidacy,[49] they also provided some of the labor that helped get him elected.[50] There is no question that the early climate achievements of the Biden administration, including investments through the Bipartisan Infrastructure Law and the IRA, would not have been possible without the efforts of numerous climate groups that worked tirelessly during the 2020 campaign cycle through the electoral process to make it possible.

In addition to participating within the institutional system, climate activists also employ outsider tactics to apply pressure on economic and political actors. These tactics range from peaceful nondisruptive demonstrations and legally permitted marches to more confrontational tactics, such as boycotts, sit-ins, and direct actions that target politics, policymakers, and businesses. In recent years, more confrontational forms of protest, including blocking traffic, throwing food, and using crazy glue to attach protesters to building entrances, roads, tarmacs, and even on furniture in TV stations, have become more common.

A BRIEF HISTORY OF CLIMATE PROTEST

The most well-known form of activism is the public demonstration. In the United States, large-scale peaceful demonstrations build on a long history of marches and protests by the civil rights, antiwar, women's rights, and environmental movements. Peaceful protest has also been a standard tactic of climate activism for decades in the United States and abroad.[51]

Since the climate regime began, the climate movement has organized large-scale demonstrations coordinated by international nongovernmental organizations (NGOs) and their local partners to take place during the climate negotiations. Some of the most notable events include: the human dike demonstration at COP6 in the Hague in 2000, the climate actions during the COP15 round of negotiations in 2009, and the shoe protest at the COP21 climate negotiations in Paris in 2015.

The human dike was the first time I ever surveyed protesters.[52] It took place during the COP6 negotiations in the Hague in 2000 where the parties were expected to finalize the Kyoto Protocol so countries could begin to ratify the treaty.[53] On the weekend between the two weeks of the negotiations, activists worked to construct a dike of sandbags around the conference hall where the negotiations were taking place. The protest was peaceful, and participants respected the boundaries set by law enforcement including UN security workers. Because it was on the weekend, many civil society groups that were participating in the negotiations as NGO observers and several other participants joined the crowd filling sandbags in the streets. I was particularly surprised to see the president of the negotiations, Dutch environment minister Jan Pronk, join the event to lay the final sandbag on the human dike in front of participants and representatives of the media from around the world.[54] The demonstration showed me how peaceful protest can be used as a tactic for groups that also work to exert pressure within the institutional system.

Three years later, in 2009, the COP15 round of climate negotiations in Copenhagen had even more protests. Like previous rounds of climate negotiations, a large-scale march was organized to take place on the weekend between the two weeks of the meetings. The march was part of an internationally coordinated global day of action around climate change. Like the human dike protest, participants in the march in Copenhagen included members of groups that were participating in the negotiations as NGO observers, local environmental groups, political parties, labor unions from around the region, and individual citizens who had heard about the event through the media.

In contrast to the human dike, however, civil society got "left out in the cold" during the second week of this round of negotiations because of increased registrations, poor planning by the Danish organizers and the United Nations Framework Convention on Climate Change (UNFCCC) secretariat, and the expansion of the climate movement to include justice-focused groups that traveled to the meeting in Copenhagen with the explicit intention of protesting it.[55] As a result, most NGO observers were blocked from participating in the final days of the negotiations. I had traveled to Copenhagen to observe how civil society groups worked as both insiders and outsiders in the negotiations to push for stronger climate action. During the weekend, I coordinated a team of researchers who surveyed participants in the peaceful march. After I was banned from participating along with other NGO observers during that second week, I sat in my cold room at the hotel and observed how groups that had planned to participate as insiders shifted their tactics and became outsiders organizing disruptive actions around the city instead.

A global day of action was also planned in 2015 to take place during the COP21 climate negotiations in Paris. Because public gatherings were prohibited in Paris during the time when the negotiations were being held after a domestic terror attack, 11,000 pairs of shoes were laid out on Place de la République on the Sunday before the negotiations began to symbolize the march that did not happen.[56] Included in the installation were shoes from Pope Francis, UN Secretary-General Ban-Ki Moon, and many others. Outside Paris, demonstrations did take place while the COP21 negotiations were underway; they involved an estimated 785,000 people from 175 different countries.[57]

As digital communication technologies have become more inexpensive and accessible, including recent advances in social media, groups have taken advantage of them, to mobilize participation in climate activism of all sorts. Like the day of action in Paris in 2015, many climate demonstrations (as well as other large-scale protest events across a range of issues) have included internationally coordinated days of action that mobilize people to participate at a flagship event in a notable location—like Paris or Washington, DC—along with locally coordinated events that are held at the same time.[58] Coordinated days of action benefit from activism that is locally embedded and coordinated by locals while being connected to similar events taking place nationally and, in some cases, internationally.

Early examples of this type of climate demonstration in the United States include the Step-It-Up day of action in April 2007 and the International Day of Climate Action in October 2009, which mobilized people from 181 countries to participate in over 5,200 actions (and was the first large-scale action coordinated by the group 350.org).[59] When I spoke with Bill McKibben in May 2023, he told me about why he started organizing these actions and cofounded 350.org to build movement capacity: "[I was] thinking, 'Well, no wonder we're losing.' We've got all the superstructure . . . [including] policy people, and we've got Al Gore . . . The only part of the movement they'd left out is the movement part. There's no people to scare or entice a politician or provide any leverage. So, we started."[60] McKibben recognized that, when local participation is coordinated with efforts at the national and international levels, these actions have more political power; They can target multiple scales of governance simultaneously, including local and domestic decision makers who determine climate policies in the specific nation-states where the demonstrations are taking place, in addition to the international climate regime.

Groups including 350.org take advantage of digital technologies, which also make it possible to mobilize civil society with much less organizational infrastructure than was needed in earlier times.[61] It is no longer necessary to plan for months to build momentum within a movement and then hold a large-scale demonstration. Instead, such events can be organized by individuals and organizations using electronic communication technologies, including various social media platforms. Innovations in communication technologies make it possible to organize large-scale protests quickly when there is widespread outrage, even if there is only limited organizational infrastructure. For example, 70 percent of participants at the 2017 Women's March in Washington, DC (which was held on the day after the inauguration of Donald Trump), reported hearing about the event from Facebook.[62]

Such innovations in digital technologies also helped to draw attention to a young Swedish activist, Greta Thunberg, who decided to strike school to draw attention to the climate crisis in August 2018. Fridays for Future was organized using digital technologies to coordinate young people who wanted to join Thunberg in her school strikes, which have come to be known as "climate strikes." Between 2018 and 2020, large-scale climate strikes were organized around the world that mobilized millions of young people and adults to draw attention to the climate crisis.[63] As of this writing, thousands of young people around the world continue to participate in climate strikes with Thunberg.

In summer 2022, I spoke with climate operative and former staffer for the Select Committee on the Climate Crisis in the House of Representatives, Aaron Huertas. We discussed how digital technologies empowered young people to get involved in climate activism at a young age. Huertas reflected on how digital technologies facilitated all sorts of activism that wasn't possible during his youth and stressed the ways digital tools were used by the youth climate movement to connect, educate, and coordinate concerned young people. "They just started . . . organizing with each other like [snaps fingers] that, right? And you can all learn from each other so quickly . . . it's been cool to see."[64]

Phil Aroneanu, who started working with McKibben while he was a student and went on to cofound 350.org before moving on to numerous other positions in progressive politics, provided a different perspective. He argued that digital infrastructure is why the climate movement was so demobilized after the 2020 election: "some of the relationships that were formed in the Trump era just haven't really continued until now. I think a lot were very light touch relationships . . . When you have mobilizations . . . millions of people are out in the streets, but it happens in the blink of an eye, the kind of slow-going relationship-building that happens around The March on Washington [in 1963] or even the People's Climate March [in 2014] that took a year to organize don't get formed. People show up, but they go back home, and then they immediately go back to what they were doing, and they're not absorbed into anything."[65]

Aroneanu notes the valuable and often overlooked role that organizational infrastructure plays in sustaining activism and channeling concern into specific campaigns. As I noted in *American Resistance*, for a high proportion of the activists who mobilized during the period of resistance to the Trump administration's policies, public demonstrations were the starting point for their activism. Organizations played an important role as the "connective tissue" for this movement, providing opportunities to connect concerned individuals to other forms of activism and engagement.[66] Without such infrastructure to channel the energy and enthusiasm of the protests in the streets into sustained activism and engagement, a movement's effects beyond the day of the actual protest are limited.

A PROFILE OF AMERICAN CLIMATE ACTIVISTS

We can learn a lot about the climate movement and how it employs insider and outsider tactics to mobilize support for climate action by looking at the profiles of participants in the most well-known forms of activism: demonstrations and marches. I have been studying who participates in large-scale climate demonstrations since the human dike protest in 2000.

Over many studies and projects, I have found that, as Huertas notes above, protests can serve as on-ramps to broader activism and engagement, including local action around elections.[67] My data also provide support for Aroneanu's point: organizations play an extremely important role in connecting activists and facilitating their transition from being a participant in a single demonstration to becoming an activist or even an organizer working for a cause.

Several large-scale protests have been coordinated in the United States to raise awareness of the climate crisis and draw attention to the need for more climate action. To date, the largest actions have been the two People's Climate Marches (PCMs), which took place in September 2014 in New York City and in April 2017 in Washington, DC.[68] In both cases, these events were flagship events of internationally coordinated days of action that included the participation of activists from 162 countries calling for action on climate change.[69]

Comparing participation at both PCMs reveals a great deal about the climate movement.[70] There is remarkable similarity between the two events in terms of the demographics of the participants, even though they took place three years apart when there were very different political contexts in the United States (one was during the Obama administration and the other took place 100 days into the Trump administration). The demographics of these crowds is consistent with my findings from progressive activism in the United States more generally during this time:[71] more women turned out at each event than men (55 percent in 2014 versus 57 percent in 2017), and the average age at both the 2014 and the 2017 marches was forty-two years old. Protest participants at the two events were also highly educated. More than three-quarters of participants at both events reported holding a bachelor's degree or higher (80 percent in 2014 and 77 percent in 2017). This rate of educational attainment among climate activists is significantly higher than the 38 percent of the American population with a BA or more as reported by the US Census in 2021.[72] These findings are not particularly surprising and are consistent with the comparative studies of climate and environmental movements across countries.[73] Although we did not collect data about the race of participants in 2014, participants in 2017 were predominantly white (77 percent). In summary, climate activists in the United States tend to be very similar to other progressive activists in recent years: they are highly educated, middle aged, predominantly white, and predominantly female.

More recently, young people have become increasingly concerned about the climate crisis.[74] As the team at the Yale Program on Climate Change Communication reported in December 2022, "While public acceptance and worry about global warming have increased over the last decade, acceptance and worry have increased faster among younger Americans aged 18–34 compared to older Americans."[75] Not surprisingly, as concern about the issue has grown, young people have also gotten more involved in climate activism.

Some young Americans were already engaged in climate activism prior to Greta Thunberg deciding to skip school in August 2018 to raise awareness about the issue, but her actions inspired many young people who had been engaged in activism around other issues to get involved in climate activism specifically.[76] As her digitally coordinated Fridays for Future spread around the world, youth climate activism and the specific tactic of the climate strike—skipping school on a weekday to protest climate inaction—became common. In March 2019, the first global climate strike took place, turning out more than 1 million people around the world.[77] Six months later in September 2019, young people and adults responded to a call by Thunberg and other young activists to participate in climate strikes as part of the Global Week for Future surrounding the UN Climate Action Summit in New York City, and the number of participants in these strikes jumped to an estimated 7.6 million people around the world.[78]

In contrast to the PCMs, which mobilized people to participate in big marches on a Saturday in one location with satellite events elsewhere, climate strikes take place on weekdays and are by design much more geographically dispersed. These protests are coordinated by young people who mobilize their friends and fellow students to strike in their communities. To understand better who participated in these events and how they compare to previous climate protests, I worked with one of the national organizers of the U.S. climate strikes—the youth-led Future Coalition—to survey the local organizers of the nationally coordinated events in 2019 and 2020. In addition, I coordinated a research team of students from the University of Maryland to survey participants at the Washington, DC–based climate strike that was part of the September 2019 global strike.

These data tell us more about who is actively participating in the climate movement and how the movement has changed as young people became more involved.[79] First, as one might expect, participation in youth-organized climate strikes in the United States involved much younger people: the median age of participants and organizers of the climate strikes in 2019 and 2020 ranged from eighteen to thirty-two years old.[80] Beyond the age difference, however, the demographics of the organizers and activists were relatively consistent with the PCMs. Participants were majority female and, while the strikes included more people of color than the PCMs, they were predominantly white.

When we look at the educational attainment of participants' parents (because many youth activists were still in school), it is clear that participants in the youth climate strikes also came from highly educated families. Around two-thirds of local organizers and strike participants reported having at least one parent with an undergraduate degree or higher. In sum, beyond their age, activists participating in the youth-led climate strikes drew from the same privileged part of the U.S. population as previous waves of the climate movement.[81]

WHAT MOTIVATED CLIMATE ACTIVISTS TO PARTICIPATE?

In addition to collecting information about demographics, I also look at what motivates individuals to participate in protest. I began collecting data about the motivations of participants during the period of the anti-Trump resistance. My research on the patterns of motivations across large-scale protests during the resistance found that the motivations that brought people out were not durable across the large-scale demonstrations during this period of heightened protest in the United States.[82]

In other words, the resistance did not have a coherent theme that held activists together and brought them out to protest in the streets again and again.[83] The climate movement, in contrast, is very different: it has been organizing for a longer period of time, and there are numerous organizations in the climate movement (including 350.org) that coordinate these large-scale events. When my collaborators and I have looked at the motivations of activists participating in climate-focused demonstrations starting at the PCM in 2017 through Earth Day 2020 at the start of the COVID pandemic, there is a lot more consistency.

Not only does my research find that climate change and the environment were consistently the top motivations at these protests, but my analysis also provides clear evidence that activists working within the climate movement have combined their concern about the climate with issues of justice and equity in the United States.[84] These findings are particularly notable because the environmental movement began with a focus on the conservation of pristine areas rather than protection from environmental harms and pollution.[85] These results show that recent efforts to merge the climate movement with the movement for environmental justice, which focuses on the unequal distribution of both environmental positives and negatives by race and class,[86] are not just window dressing by national organizations.

Although the people participating in the climate movement continue to be highly educated and predominantly white, racial justice and equality have become common motivations among participants in this movement. Environmental organizers and activists report many reasons beyond climate change and the environment to participate in demonstrations, and these patterns of motivations are notably durable over multiple years of climate activism. In other words, these findings from protest participants at climate-focused events show that the connection between the climate and climate justice movements is not just symbolic signaling by organizational leaders. Although the identities of the people protesting continue to be less racially diverse and relatively privileged, the activists joining the protests in the streets consider issues related to climate, equity, and racial justice as main motivations for their activism.[87]

WHAT ELSE DID CLIMATE ACTIVISTS DO?

My surveys also ask climate activists about other ways they had participated in politics over the past year (see table 3.1). The responses showed that those who engaged in climate protests and strikes are not disengaged from the political process. In fact, most of them report participating in a range of civic activities—including contacting elected officials, attending town hall meetings, and communicating with the media to express a view. In addition, most activists at all the events except the DC climate strike in September 2019 reported expressing their political and environmental views as individual consumers by boycotting and/or buycotting products. Taken together, it is clear that the people participating in climate activism are not new to activism. Rather, they are what political scientists would call joiners who have integrated protesting into their civic lives.[88]

TABLE 3-1 Civic engagement of climate activists over time

In the past year, have you . . .	PCM 4/2017 (%)	Organizers of 9/19 Strikes (%)	DC Strike Participants 9/19 (%)	Earth Day Live 4/20 (%)
Contacted an elected official	70	83	46	85
Attended a town hall meeting	58	77	46	79
Contacted the media to express a view	49	64	17	63
Boycotted or deliberately bought certain products for political, ethical, or environmental reasons	76	93	44	77
Participated in direct action	41	61	32	69

In addition to participating in more insider forms of engagement through the state or the market, activists also reported participating in outsider tactics, including direct action. Direct action incorporates civil disobedience that can be either peaceful or more confrontational and aggressive.[89] About a third of the activists who were surveyed while participating in the 2017 People's Climate March and the 2019 climate strike reported participating in direct action in the past twelve months. In addition, over 60 percent of the local hosts who were organizing the September 2019 climate strikes around the United States and the Earth Day Live events in 2020 also reported having participated in some sort of direct action in the past year.

More recently, Earth Day 2023 protests took place around the United States. In Washington, DC, a small crowd commemorated the day and called for the Biden administration to end the era of fossil fuels. Participants at this event were also predominantly female, white, highly educated, and relatively young (mean age of thirty-three years old), with a number of them reporting that they were affiliated with the youth-led Fridays for Future. Fifty-one percent of the participants in Earth Day 2023 reported participating in direct action in the past year.

That climate protest participants report various forms of engagement is not unique to the climate movement. Across protests about a range of progressive issues during the period of the anti-Trump resistance, I found activists to be highly civically engaged.[90] In many ways, these findings echo the conclusions of a 2022 report by the Deloitte Center for Integrated Research, which documents the drivers of "personal sustainability" across twenty-four countries.[91] In the report, the authors note how individual and collective political and economic behaviors are connected for the most engaged, whom they call "sustainability standard-setters." Like the data I collected from climate activists show, this report highlights that standard-setters connect their individual behaviors to collective efforts that can have both direct and indirect effects on climate change.

In the next chapter, I provide more detail about how climate activists are employing a more diverse range of tactics than ever before. They are bundling insider and outsider tactics, including civil disobedience, to achieve their goals and mobilize the masses to take action against the climate crisis.

Chapter Four

SAVING OURSELVES WON'T BE POPULAR AND IT WILL BE DISRUPTIVE

ON THE second week of 2023, Greta Thunberg traveled to Lützerath, Germany, to participate in a protest against the demolition of a village that would make way for a coal mine.[1] In contrast to the peaceful and nonconfrontational marches, speeches, and demonstrations she had become known for since she started skipping school in 2018, Thunberg joined activists to engage in nonviolent civil disobedience. The action involved occupying the village with the goal of stopping its destruction. By the end of the weekend, all the activists—including Thunberg—had been detained by law enforcement. Videos of police kettling protesters and struggling in the mud were all over social media, and the town was razed to the ground.

What happened in Germany was notable for two reasons. First, it highlights the fact that even countries with a reputation as so-called climate leaders are continuing to extract fossil fuels and expand their capacity to burn them.[2] Second, the protest in Lützerath shows how one of the most well-known youth climate activists has expanded her activism to include more confrontational tactics. By participating in nonviolent civil disobedience in Germany and being detained for doing so, Thunberg joins a growing number of people in the climate movement who are frustrated with the incremental political responses to the climate crisis that will not achieve the systemic changes required to save ourselves.

Thunberg explained the growing urgency for more activism during an event to launch her climate book in October 2022: "There is simply no other option. . . . We have to abandon the idea that the people in power will come to our rescue. . . . The people in power have proved that they are not going to be the ones leading this change—not without massive public pressure from the outside. We have to be that pressure. . . . Some people are taking action . . . but we need to be many more."[3] There is no doubt that this decision by one of the most well-known climate activists in the world to expand her tactics and engage in civil disobedience will inspire members of the climate movement—both young and old—to get more confrontational. Thunberg continued to engage in nonviolent civil disobedience throughout 2023. We should interpret this decision as a harbinger of what we can expect from climate activists in the years to come.

In the days following the Lützerath protests, the internet was atwitter with hot takes on Thunberg's decision to integrate more confrontational tactics into her personal activist's toolbox. During his comments at the World Economic Forum just days after the German town was razed, former U.S. vice president Al Gore took time to praise Thunberg's actions. His comments started with his usual technocratic tone, but they became increasingly impassioned as he spoke about how fossil fuel interests have captured the policymaking process while youth climate activists grow more concerned. "Emissions are still going up. . . . We have to act. . . . Greta Thunberg was just arrested in Germany. I agree with her efforts to stop the coal mine in Germany. . . . There are some meaningful commitments, but we are still failing badly."[4] This statement from the former vice president and Democratic presidential candidate highlights how new battle lines are being drawn around the climate crisis and what are considered to be acceptable responses.

This chapter focuses specifically on the broader array of tactics now being used by the climate movement. I begin by discussing the rift in the climate movement that was exacerbated by the passage of the Inflation Reduction Act (IRA). Since its passage in summer 2022, we have seen a growing radical flank emerge in the climate movement that is pushing for more confrontational activism. After describing this radical flank and what we can expect from it, I disaggregate the different types of direct action that are becoming more common from the "shockers" and the "disruptors" to understand their variable focus and likely outcomes. The final section of the chapter summarizes some important lessons learned from the ways civil disobedience was integrated into the struggle for racial justice in the United States. Building on these findings, I discuss how the expanded range of tactics can contribute to the fight to save ourselves.

HOW INCREMENTAL POLITICS EXACERBATED THE GROWING RIFT IN THE CLIMATE MOVEMENT

As I already discussed, climate successes achieved during the first half of the Biden administration were won thanks in part to the efforts of the climate movement. Passing the IRA involved a significant amount of compromise from the Biden administration and Democrats in Congress. These compromises included concessions to fossil fuel interests. As the bill became weaker and less threatening to those invested in the fossil fuel industry, it also became more likely to pass. At the same time, the compromises that policymakers made to get the bill over the finish line put substantial stress on the climate movement and split support for specific efforts, actions, and tactics.

The battle over the incremental compromises that made passage of this reconciliation package possible turned off numerous climate groups—particularly groups that prioritize environmental justice, which in turn includes many youth-led organizations. These groups had toiled throughout the four years of the Trump presidency to mobilize support for climate action that prioritized what they called a *just transition*, which stresses "the idea that justice and equity must form an integral part of the transition towards a low-carbon world."[5] In other words, they prioritize combining climate justice with racial and economic justice, which is consistent with the priorities of the climate activists whom I have studied at protests since 2017.[6]

As negotiations went on and the bill became weaker, it also became clear that the final version of the IRA would not follow through on the administration's environmental justice–related promises to prioritize protecting frontline communities.[7] Instead, it won the support of political moderates by providing opportunities to expand fossil fuel infrastructure that was expected to be sited in or near these communities.[8] As a result, the pared-down IRA was criticized by many environmental justice–focused groups that stressed the ways that the bill did not invest sufficiently in communities of color and Indigenous communities who are the front line of the climate crisis. In his summary of the IRA, racial and climate justice advocate Anthony Rogers-Wright called it a "poison pill for Black and Indigenous communities and movements."[9]

The conflict among climate groups regarding the IRA is quite different from the ways progressive groups worked together as a unified coalition over the four years of the Trump administration. Climate groups were an active part of the resistance coalition, working in collaboration with other left-leaning organizations as part of the American resistance to support the Democratic agenda and convert outrage over the Trump administration and its policies into resistance in the streets and political power through the electoral process.[10] The resistance organized various large-scale peaceful protests over this time period, which included the climate-focused March for Science and People's Climate March (PCM), as well as the multiple women's marches (including the first one that took place on the day after Trump's inauguration in 2017), and the March for Our Lives.[11]

Two years into the Trump administration's term, the *Washington Post* reported that one in five Americans had participated in a rally or protest between 2016 and 2018.[12] As I discuss in detail in *American Resistance*, left-leaning Americans participated civically and politically at very high levels during this period. In chapter 3, I presented data showing how climate activists, many of whom were part of the resistance, had similarly high levels of engagement.[13] These highly engaged Americans joined civic groups that maintained a durable coalition of progressive interests and worked together to respond to the Trump administration and its policies.[14] Throughout the four years of the Trump administration, most left-leaning groups worked hard to ensure that this coalition held and that their activism stayed peaceful with minimal confrontation. These efforts continued through the summer of 2020, when protests broke out across the United States in response to the police killing of unarmed George Floyd in Minnesota.[15]

As this period of heightened progressive activism with limited confrontation wore on, journalists questioned why left-leaning groups were not engaging in more mass mobilizations and why the protests weren't more disruptive. Sometimes they publicly challenged the decisions of resistance groups to focus their efforts on effecting social change through the institutional political system and elections instead of employing more outsider tactics.[16] Nevertheless, the diverse resistance coalition, which included activists focused on democratic reforms, stopping gun violence, ending systemic racism, and limiting climate change, held because it saw the Trump administration and its policies as a common enemy and our democratic elections as the solutions.[17] While the coalition focused on institutional political change, groups on the political right became increasingly confrontational, with armed protesters occupying statehouses, marching in the streets, and even attacking the U.S. Capitol on January 6, 2021.[18]

There is no question that the resistance coalition played a substantial role in in the 2020 election and helped Joseph Biden win the presidency. With Biden moving into the White House and the Democrats taking control of the Senate, many of the activists who had resisted throughout the Trump administration, working tirelessly to get Democrats elected, were hopeful that they could stand down and let the Democrats whom they had elected follow through on their promises for progressive reform. It is not uncommon for progressive movements to demobilize when Democrats win and activists stand back and let the people they elected serve.[19] A year into the Biden presidency, it was clear that the anti-Trump resistance followed a similar pattern, struggling to pivot from defense to offense and maintain a broad level of engagement while working to make progressive campaign promises a political reality.[20]

In summer 2022, I checked in with a long-time organizer specifically about what had happened to the resistance coalition after the 2020 election. She responded: "The progressive and liberal coalition that got Trump unelected . . . that's started to fall apart . . . We're in a moment where we have to basically start from scratch . . . [I'm surprised] how demobilizing COVID and the Biden presidency has been even among progressives . . . The passage of the Inflation Reduction Act happened in some ways *despite* that demobilization."[21]

As it becomes increasingly apparent that the incremental politics coming out of the political system—including the IRA—will not put us on a path to the systemic changes needed to respond to the climate crisis, a clear rift has formed within the civil society sector. The more professionalized environmental groups,[22] which are sometimes referred to as the Big Greens, with paid staff, large budgets, and offices in Washington, DC, have labored tirelessly to encourage climate action by taking advantage of the institutional political levers that exist for engaging with businesses and members of the government inside the system. After the passage of the IRA, these groups have been relatively quiet as the administration opens more public lands to fossil fuel extraction.[23] In contrast, the contingent of the climate movement that has focused on responding to climate change with a just transition finds supporting Biden after his compromises and fossil fuel giveaways to be a bridge too far.[24]

ENTER THE RADICAL FLANK

While most Big Green groups were celebrating (and supporting) the IRA, even though it did not address so many critical issues related to climate justice, many other groups started pulling away from the broader climate coalition. They determined that more confrontational activism was needed to get beyond incremental politics to the necessary systemic changes. These activists had come to similar conclusions as Thunberg: civil disobedience and direct action were necessary to pressure governments and businesses to act at the level required to respond to the climate crisis.

Disagreement among activists and organizations involved in a social movement is not unique to the climate movement:

internal conflicts have been well documented across a range of movements, including the civil rights movement;[25] the women's movement;[26] and the lesbian, gay, bisexual, transgender, queer/questioning (LGBTQ) movement.[27] Conflicts within movements sow the seeds of dissent and cultivate what scholars call a "radical flank."[28] In their 2022 paper, Brent Simpson, Robb Willer, and Matthew Fenberg define the radical flank as "a discrete activist group within a larger movement that adopts an agenda and/or uses tactics that are perceptibly more radical than other groups within the movement."[29] A radical flank emerged in the civil rights movement, and the tactics became more confrontational after activists realized that they did not have the necessary access to power to make change through the legal and political systems.[30] Civil rights activists employed more disruptive tactics, including staged sit-ins, nonpermitted marches, and even riots. In most cases, these tactics were initiated by youth-led organizations,[31] including some with rather militant ideas. Radical flanks are common when there is conflict over tactics, targets, timetables for action, or perspectives on a policy (like the IRA).

As a radical flank has grown in the climate movement, more and more groups have followed the example set by activist and organizer Bill McKibben, who began organizing acts of civil disobedience to stop the Keystone XL pipeline in 2011. When we spoke in May 2023, he explained his decision to expand the tactics used during the pipeline protests. He had concluded that climate activism would have more impact if it was "confrontational as well as educational." He continued: "I decided we should try . . . asking people to come to Washington and get arrested . . . it really just blew up. It turned into the largest civil disobedience action about anything in this country in a long time . . . It was very effective there and kind of galvanized the movement."[32] Since 2011, McKibben has been a prominent activist in the growing radical flank of the climate movement.[33]

Although he is one of the most recognized, McKibben is not alone in his support of more confrontational climate activism. I sat down with the executive director of the U.S. Climate Action Network (USCAN), Keya Chatterjee, in August 2022 as the IRA was just being finalized. She explained the perspective of USCAN, which includes more than 190 groups that are working to fight climate change in a just and equitable way.[34] During our conversation, Chatterjee summarized what the group had learned from the negotiations around the climate bill that had evolved from a campaign promise to the IRA. "What this bill exposed is that the system itself is too deeply flawed to address this crisis." She also noted how other green groups had a different take on the IRA: "There are a lot of institutionalists in our line of work who just will cling to institutions even when they're not serving us."[35]

Groups including USCAN were frustrated by all the efforts and resources being invested in incremental policies like the IRA that will neither protect frontline communities nor achieve the more substantial systemic changes that are needed. Chatterjee also discussed how funding for environmental groups is exacerbating the problem: "Stop giving big groups money. They literally are burning it, and they will let the planet burn as they continue to burn the money. Stop giving them money. Stop funding reformer tactics."

While Chatterjee was particularly frank in her critique of Big Green groups and the system that supported them, many other climate- and justice-focused groups were less blunt. One such group was the youth-led Sunrise Movement, which gained notoriety when they occupied then Speaker of the House Nancy Pelosi's office with newly elected congressional representative Alexandria Ocasio-Cortez to send a message about the climate crisis and the need for a Green New Deal.[36] In her email to members in April 2023, cofounder and leader of the Sunrise Movement Varshini Prakash summarized their accomplishments: "We delivered a mandate to the Democratic establishment that, to garner the votes of our generation, they must commit and deliver on a climate plan in line with what science and justice demand. . . . Two years later, the climate movement won the largest national climate bill in world history—which, while flawed and inadequate, marked the first time the federal government had acted on climate in history."[37] Although Prakash's message in April 2023 acknowledged the incremental progress of recent policies, the group was mobilizing activists by stressing what it considered the weaknesses in the IRA. The welcome message on the Sunrise Movement website in summer 2023 used aggressive language to describe its plans: "We will force the government to end the reign of fossil fuel elites, invest in Black, brown and working class communities, and create millions of good union jobs. We're on a mission to put everyday people back in charge and build a world that works for all of us, now and for generations to come."[38]

While groups like the Sunrise Movement worked to mobilize more people to join a movement that focuses on a just transition, the level of progressive activism was much lower than during the glory days of the resistance to the Trump administration. In my conversation with an organizer, who had worked in a range of positions in the climate movement for over twenty years, she reflected on the climate activism during the first two years of the Biden administration: "There have been some [recent] attempts at mobilization and they were pretty sad, to be honest."[39] Consistent with this organizer's perspective and with the research that finds that Democratic gains can take the wind out of the sails of left-leaning movements, progressive activism, including climate activism, had gone down substantially since the Biden administration took office in January 2021.[40]

Most progressives had stayed home and waited to see what the new president would accomplish, even as the Supreme Court issued decisions limiting reproductive rights and the Environmental Protection Agency's (EPA) authority to regulate power plant emissions. During this time, the radical flank of the climate movement grew and coordinated smaller actions around the United States. These groups focused on centering climate justice and limiting fossil fuel expansion. The actions they coordinated were more confrontational, in many cases involving nonviolent civil disobedience.

One such action took place on Earth Day 2023 in Washington, DC. The event included a rally and a march. The local DC branch of Extinction Rebellion (XR) had also planned a direct action to take place during the event, which was rumored to involve a die-in on Pennsylvania Avenue. The action did not happen, however, because of severe weather. Before the storm rolled in, my students and I surveyed a sample of 143 activists from the crowd of more than 1,000 participants as they prepared for the march and rallied. The findings from this survey provide evidence regarding how the protesting population had changed since the 2020 election. One of the most notable findings from Earth Day 2023 is the fact that the Big Greens were not involved in mobilizing participants to join the activism in the streets. Even though a third of participants reported hearing about the event from an organization, the groups named were almost exclusively groups from the radical flank of the climate movement. The most common organizations that participants mentioned were Extinction Rebellion, Fridays for Future, the Rachel Carson Council, Sunrise Movement, and Third Act.[41]

LEANING INTO CONFRONTATIONAL DIRECT ACTION

Although a small number of groups have been engaging in civil disobedience for years, the tactic has become much more common. Perhaps the most well-known group that has been pushing back against incremental climate politics and using direct action is XR. Since it began in the United Kingdom in 2018, the group has spread to eighty-eight countries. In 2023, it included 1,029 groups in its decentralized network of activists.[42] XR aims to disrupt business as usual and get arrested doing so to draw attention to the climate crisis. In November 2021, the group posted a thread on Twitter (renamed X in July 2023) that made the case for joining the growing radical flank of the climate movement. It explained why incrementalism is insufficient to meet the challenge of the climate crisis:

> There has long been a tension between incrementalists in the environmental movement, and people like us who demand radical change . . . It would be great if we could stop the breakdown of climate and nature through existing power structures and incremental changes. System change is scary and hard to bring about . . . But we have to accept that incrementalism has failed for the past 30 years, and it's too late now . . . Don't like the sound of it? Would you prefer societal collapse? Above all, it's a moral question. How many cities, countries, people and species will we allow to be destroyed before we say, "Enough is enough. We demand action"? . . . Are you ready to say, "Enough is enough"? Please join us.[43]

When I spoke with leaders of groups that were engaging in civil disobedience in the United States—some of which are directly connected with the global XR network—many of them echoed this perspective. For example, a spokesperson for XR NYC provided a critique of mainstream progressive activism in the United States and noted the impotence of large-scale peaceful demonstrations that had been the hallmark of the anti-Trump resistance. "Most of the time . . . a coalition of groups will get together, they'll have a list of demands, and it doesn't really get executed . . . It's more performative, and it just happens in one day, and then after that, everybody just goes home."[44]

Several other activists had similar perspectives. For example, a leader of a different group told me about her journey to disruptive climate activism: "It was the Trump years that radicalized me . . . I was one of those super activists—went to everything, started organizing, did my first civil disobedience around the Kavanaugh hearings [when Brett Kavanaugh was being confirmed to sit on the U.S. Supreme Court], which was a big deal for me at the time . . . That led me to believe that civil disobedience—disruption—is the only thing that even can come close to working."[45]

I expressed a similar frustration with the lack of follow-through by left-leaning groups and their reluctance to engage in more confrontational activism in a piece I published in *The Nation* with Vanessa Williamson in summer 2022: "It's Time for Democrats to Stop 'Clapping for Tinkerbell.'" We were motivated to write the article after observing how left-leaning groups and elected officials wrung their hands rather than mobilizing the masses after the Supreme Court decision that overturned *Roe v. Wade* was leaked to the public. Our piece noted that progressives have been following "an outdated playbook of one-day rallies and electoral politics that currently can achieve no more than a pro forma vote on doomed federal legislation." Calling for a peaceful march and some lobbying days to try to get Congress to pass a bill redirects civic outrage into symbolic and performative measures that can backfire on a social movement.

Although the leaked Supreme Court decision about their plans to overturn *Roe v. Wade* motivated our article, we stressed how these challenges were consistent across a range of issues, including gun reform and climate change. The piece concludes: "If they close the doors on confrontational activism and civil disobedience, mainstream liberal and Democratic organizations cede a whole range of demonstrably effective tactics to their opponents."[46] This asymmetry could be seen clearly as most of the political left had continued to push for insider tactics that might yield incremental gains while the political right grew increasingly confrontational. During this period of time, numerous right-leaning groups had organized protests that threatened violence in response to various COVID policies. Moreover, on January 6, 2021, right-leaning groups who supported President Trump staying in office after he lost the 2020 Presidential election attempted an insurrection at the U.S. Capitol that involved clashes with Capitol police, vandalizing and looting of the Capitol building, as well as activists building gallows on the grounds of the U.S. Capitol, which they claimed were constructed to hang Vice President Pence and other elected officials.[47] With numerous conservative groups embracing tactics that elicit fear and employ violence to achieve their goals, the Republican Party and its leadership was reluctant to respond with criticism.[48]

As the climate crisis worsens and more and more concerned activists lose confidence that institutional politics can adequately address the problem, the radical flank will grow. In spring 2023, on the heels of the IRA success, the Biden administration followed through with its all-of-the-above energy plan that involved expanding fossil fuel infrastructure while funding an energy transition to more clean and renewable sources. The administration approved the Willow Project to drill for oil on public lands in Alaska, and more climate groups lost faith. Before these decisions, millions of young activists had taken to TikTok to protest the Willow Project, which opened a huge area of Alaskan wilderness to drilling in March 2023. Even though the online mobilization to #StopWillow started trending (marking the first time a climate-focused hashtag ever had trended on TikTok) and concerned activists contributed over 1 million letters and millions of signatures to an online petition to stop the project, it was approved.[49] On the day of its decision, Gen-Z for Change, which was one of the groups organizing the activism on TikTok, called the approval of the project "a slap in the face to young people across the country."[50]

Gen-Z for Change was one of many environmental and climate groups that had joined the president and congressional leaders to celebrate the passage of the IRA at the White House in September before the 2022 midterm elections. At the celebration, James Taylor sang to a crowd of thousands that included representatives from Big Green groups and youth climate organizations, along with numerous climate influencers who shared images and videos from the event with their millions of social media followers.[51] In the wake of the Willow decision, many activists reported feeling duped by the president. Elise Joshi of Gen-Z for Change posted: "He's not going to be able to sweep this under the rug like he might have been able to do a month ago. . . . People are watching and people do not want him to do this."[52] The Willow Project was just the first in a series of big projects that the Biden administration approved to expand fossil fuel extraction in the United States. In July 2023, Joshi followed through and attended an event where the White House Press Secretary was speaking and interrupted: "Excuse me for interrupting, but asking nicely hasn't worked out. A million young people wrote to the administration pleading not to approve a disastrous oil drilling project in

Alaska, and we were ignored. So I'm here channeling the strength of my ancestors and generation."[53]

As fossil fuel expansion continues in the United States despite the administration's stated commitment to a clean energy transition, it is very likely that activists who have utilized insider tactics, as well as mobilizing against these projects by expressing their criticism online will join Joshi and many others to redirect their activism and join the radical flank of the climate movement.

USING DIRECT ACTION TO SHOCK AND DISRUPT

As I already noted, civil disobedience is not new to the climate movement, but tactics that involve direct action have become much more popular in the past few years. Direct action involves disruption and civil disobedience, including nonviolent and violent tactics like sit-ins, general strikes, vandalism, monkeywrenching, and even riots.[54] Direct action is being used to achieve a range of climate-related goals by a growing list of groups. These more confrontational efforts by climate activists fall into two main categories: direct action to elicit shock and gain public and media attention, and direct action to disrupt as part of a broader campaign.

DIRECT ACTION TO ELICIT SHOCK

On October 14, 2022, two young climate activists walked into the National Gallery in London, opened up a can of tomato soup, threw the soup on a painting of sunflowers by Vincent Van Gogh (which was covered in a protective coating), and then crazy-glued their hands and stuck themselves to the wall below the painting's frame while giving a speech about fossil fuels and the energy crisis in the United Kingdom.[55] This action was one of many in fall 2022 that were part of the Fall Uprising. The uprising involved a coalition of groups that organized sustained disruptive protest throughout eleven countries during the month before the COP27 round of climate negotiations began in Egypt.[56] It included activists blocking traffic, interrupting talk shows, and throwing food at famous works of art as a way to gain attention for the climate crisis and the need for more aggressive climate action.[57] The Fall Uprising was funded by the Climate Emergency Fund, which explained the motivation for supporting the shockers and their activism on their website: "We support the organizations who tell the truth, demand transformation at emergency speed, and put everything on the line to protect humanity and the living world."[58]

The main goal of this type of activism is to raise awareness about the climate crisis by using acts of civil disobedience to gain media attention. Frequently, this type of action also elicits a lot of anger and criticism from the general population.[59] One group embracing this strategy in the United States is Declare Emergency, which is part of the international A22 Network and describes itself as "a campaign using nonviolent civil resistance techniques to disrupt the status quo and demand that our government take meaningful action to address the climate emergency."[60] A volunteer organizer with the group explained the network and their tactics to me when we spoke in March 2023: "we share strategy and tactics of nonviolent but highly-disruptive actions, calling for climate action by our governments . . . [Our type of activism expects] the majority of people participating in an action to risk arrest . . . which disrupt[s] the general public, instead of focusing on an industry or political target."[61] This group is one of many in the United States that follows the example set by XR, which focuses on nonviolent civil disobedience designed to end in arrest. In their book about XR and climate activism, Oscar Berglund and Daniel Schmidt explain: "Arrests were not just an inevitable result of the disruption, but an end in itself."[62]

In an article for the *Guardian*, members of the United Kingdom–based group Insulate Britain, which blocked roads in the United Kingdom in 2021, explained why they had chosen this type of direct action: "'We managed to keep going out for so long and still get the media attention . . . Even if people thought it was negative, it planted a seed.' The only way to get attention was to cause serious disruption. Holding placards and signing petitions was not going to cut it."[63] This quote by an activist from Insulate Britain summarizes the general sentiment among many shockers: their civil disobedience is choreographed explicitly to attract attention from the general public and the mass media.

A similar explanation was provided by actress and activist Jane Fonda when she described her Fire Drill Fridays. Inspired by Thunberg and her Fridays for Future, Fonda started these actions in fall 2019. They involved acts of civil disobedience that were intentionally designed to lead to the activist's arrest and to draw attention to the climate crisis. The weekly actions brought together notable people, including "celebrities, youth, Indigenous leaders, representatives from impacted and under-represented communities, as well as movement and thought leaders" to participate in "weekly protests centered around civil disobedience and a demand [that] Congress pass the Green New Deal."[64] In an interview, Fonda explained her activism: "Civil disobedience is not a first resort, but it's a step up. You've petitioned, marched, pleaded, and begged, and you haven't been heard, so you take the next step. To align your body with your values is very empowering, and this offers that opportunity."[65] As a working actress who has won numerous awards, Fonda was able to take advantage of her platform and notoriety to gain substantial media attention for her acts of civil disobedience and subsequent arrests.[66]

Another group that follows this model of engaging in direct action to shock the public and get media attention is Scientist Rebellion. The group is known for its activists wearing lab coats at protests and is comprised of "scientists and academics who believe we should expose the reality and severity of the climate and ecological emergency by engaging in non-violent civil disobedience . . . We are terrified by what we see, and believe it is both vital and right to express our fears openly."[67] Climate scientist Rose Abramoff wrote about her experience with the group protesting at the annual meeting of the American Geophysical Union (AGU) in the *New York Times* in January 2023. "My fellow climate scientist Peter Kalmus and I unfurled a banner that read 'out of the lab and into the streets.' In the few seconds before the banner was ripped from our hands, we implored our colleagues to use their leverage as scientists to wake the public up to the dying planet."[68] The protest lasted a very short time between plenary speakers before security at the AGU meeting escorted the two scientists out of the meeting and a professional misconduct inquiry was initiated. As news spread about the incident, Abramoff was fired from her job working as a scientist for the Oak Ridge National Laboratory. I spoke with her on the phone two days after her piece was published in the *New York Times* to learn about why she had chosen this form of activism. She explained to me that she considers civil disobedience to be "severely understaffed" as a tactic in the climate movement and she is "well-equipped to do it."[69]

Shockers were also active in April and May 2023 in Europe and North America. In the United States, they stopped traffic, blocked the entrance to the White House Correspondents' dinner, and smeared paint on the casing around Degas's *Little Dancer of Fourteen Years* sculpture in the National Gallery in Washington, DC. In response to the action at the National Gallery, which was organized by Declare Emergency, the editorial board of the *Washington Post* published a scathing editorial, calling the action "counterproductive": "This kind of 'protest' is no protest at all. It is vandalism plain and simple, and, perhaps more than anything, it harms the cause these 'protesters' claim to care so much about."[70]

Although this editorial was one of many that claimed this type of activism was detrimental to the broader climate movement, the evidence to date does not support this perspective. In an article published during the Fall Uprising in 2022, social psychologist Colin Davis summarized the general social science research on the effects of this type of activism. He concludes that disruption "may actually be a very effective way to increase recruitment . . . The existence of a radical flank also seems to increase support for more moderate factions of a social movement, by making these factions appear less radical."[71]

These findings are consistent with research conducted on the radical flank in other social movements. For example, in their study of the radical flank in the women's movement in the 1950s through the 1970s, Holly J. McCammon, Erin M. Bergner, and Sandra C. Arch find that this type of within-movement conflict can create opportunities to achieve the movement's broader political goals.[72] In his research on disruptive tactics during the civil rights movement, Doug McAdam finds that, by employing innovative tactics, activists created moments of "increased bargaining leverage" for the movement.[73]

These findings are also corroborated by data collected from the general population. In a 2022 survey conducted by the Yale Program on Climate Change Communication, over a quarter of the U.S. population was found to support generally "an organization engaging in non-violent civil disobedience (e.g., sit-ins, blockades, or trespassing) against corporate or government activities that make global warming worse." When the data were broken down by political orientation, 56 percent of the people who identified as liberal Democrats showed support for an organization that engages in nonviolent civil disobedience.[74] It is important to note that these findings are based on general opinions; they are not actual responses to a particular act of civil disobedience with a particular target. Although much more research is needed to understand the broader effects of specific actions on public opinion, political opportunity, and media coverage,[75] most of the shockers and the groups that fund them use media coverage—such as the number of sources covering the story or impressions on social media—as the main indicator of the success of this type of activism. When we look at the media coverage of direct action to elicit shock, there is little question of the success: the shockers and their actions have drawn substantial attention (and media coverage) to the issue.

DIRECT ACTION TO DISRUPT AS PART OF A BROADER CAMPAIGN

In contrast to the shockers who use direct action to gain attention through the media, other activists integrate direct action into what scholars of social movements refer to as their broader "repertoire of contention" to address the climate crisis.[76] Youth-led climate groups have become well known in recent years for employing this type of nonviolent civil disobedience as part of their climate campaigns. One example is the Sunrise Movement, which was started by several activists who originally cut their teeth in the divestment movement on college campuses and is recognized for its fall 2018 occupation of Nancy Pelosi's office when she was serving as Speaker of the House of Representatives. In her email to members before stepping down as leader of the group five years later, Varshini Prakash summarized the ways the group had combined direct action into their broader campaigns: "In the months and years after [the sit-in in Pelosi's office], we fought to keep climate at the center of American politics through relentless sit-ins, office visits, media presence, and political engagement."[77] Many of the local chapters of the Sunrise Movement, which they call hubs, have used this same model of employing diverse tactics to pressure their universities to divest from fossil fuels. Along with efforts to lobby university leaders and alumni, these groups also use direct action—such as occupying buildings and dropping banners—as part of their activism to gain more attention for their efforts.

The Sunrise Movement is not the only youth-led climate group that has integrated direct action into its broader campaigns. Perhaps one of the most well known of these groups is Fossil Fuel Divest Harvard, which coordinated activism from students, alumni, and faculty to pressure Harvard to divest from fossil fuels. Before the university agreed to divest its endowment in 2021, the group had been integrating civil disobedience and direct action into their activism for years.[78] In addition to working to build support from famous alumni and lobbying the university administration, the group also coordinated disruptive actions, including occupying buildings and even interrupting the Harvard-Yale football game by storming the field during halftime in 2019.[79] Ilana Cohen, who was one of the primary student organizers of the action during the Harvard-Yale game, reflected on the group's decision to integrate civil disobedience into their tactics. In a conversation with me right before she graduated from Harvard in 2023, Cohen explained that storming the field during the game was "part of a much broader campaign . . . It helped create the conditions in which we could effectively pursue actions like legal complaints that further challenged our universities' complicity in climate injustice. It worked in conjunction with those tactics to make anything other than a divestment commitment unacceptable."[80]

Direct action has also been integrated in other youth-led campaigns. When I sat down with youth climate leader Jacob Lowe in December 2022, he told me about how he started organizing when his school's Sunrise hub at George Washington University had won their divestment campaign. In the follow-up to that campaign, Lowe and other student activists redirected their energies into working to stop fossil fuel companies from investing in university research. These efforts were later formalized as the Fossil Free Research campaign, which puts pressure on universities to refuse research funding from the fossil fuel industry. In a public letter to universities, the group states, "We believe this funding represents an inherent conflict of interest, is antithetical to universities' core academic and social values, and supports industry greenwashing."[81]

Lowe spoke with me about the tactics that students have been using in the divestment and Fossil Free Research campaigns. He explained how student groups are combining disruptive tactics with more institutional ones to raise awareness and mobilize new members to join. "When you do that kind of thing right, you get attention. It's the attention from people who wouldn't have seen the work you're doing otherwise." Lowe continued, "I think disruptive actions can go a long way toward also getting people excited to get involved . . . It's a way to build our base and get more people plugged in . . . Anything we can do to get this idea out into a popular discourse goes a long way."[82]

This type of direct action is not limited to youth-led activism. In 2023, Bill McKibben launched a new group called Third Act. The group's website explains that it is "a community of experienced Americans over the age of sixty determined to change the world for the better."[83] In addition to mobilizing members to take personal actions like moving their money out of banks that invest in fossil fuels, switching their credit cards, and lobbying financial institutions and communicating with others in their social networks to expand participation, Third Act also coordinates days of nonviolent civil disobedience. One of their first acts of civil disobedience was called the *Rocking Chair Rebellion* and involved activists blockading the entrance to banks with hand-painted rocking chairs in one hundred locations across the United States in March 2023.[84] Painted rocking chairs were back in the streets the following month at numerous climate actions in Washington, DC, including Earth Day 2023 and at the activist blockade at the White House Correspondents' dinner a week later.

Leaders of other climate-focused groups have also decided to broaden their activism to include civil disobedience as part of their tactics. One such leader is Rabbi Jennie Rosenn, the founder of Dayenu: "a movement of American Jews confronting the climate crisis with spiritual audacity and bold political action."[85] In an email to the group's list, Rosenn shared her decision to participate in civil disobedience and get arrested with other spiritual leaders as part of the People vs Fossil Fuels week of action in October 2021. "I believe it is our moral imperative to do everything in our power to address the climate crisis, and the magnitude of the crisis demands that all of us step up, and step outside of our comfort zone. For me, this week, that meant risking arrest, in an act of civil disobedience."[86]

When I asked Rosenn if she would get arrested again as part of her climate activism (and recruit members to join her), she told me: "I would absolutely participate in civil disobedience again . . . It builds power and is bonding for the activists participating, which is important in its own right, but/and it's also critical that it be done in a way, at a time, and with a target that could truly bring about meaningful change." Rosenn went on to explain how disruption should be integrated into campaigns. "The climate movement should absolutely continue to engage in civil disobedience at the moments and in the situations where this tactic is assessed to have real potential to catalyze the necessary change . . . [it] is an important part of the mosaic of social change, activism, and movement building."[87]

In many ways, Rosenn's words echo the message from McKibben when he spoke to me about how disruptors embed their civil disobedience into a broader campaign. The cofounder of 350.org, founder of Third Act, and organizer for the civil disobedience against the Keystone XL pipeline told me: "People have to figure out how to do it, how exactly it ties into some particular fight that you're engaged in . . . Because obviously, your target's not the art museum. Your target's an oil company or a politician or something, and they all know that."[88] It's important to stress that McKibben, a long-time activist, is not being critical of the shockers and their tactics. He added: "I'm generally supportive of these protests and do not think they've turned off more people than they attracted; I'm a supporter."[89]

LEARNING FROM THE CIVIL RIGHTS MOVEMENT

When we look back at the U.S. civil rights movement, there are lessons to be learned about how to integrate civil disobedience into what Rosenn called "the mosaic of social change." In his research on the diffusion of confrontational activism as part of the civil rights movement from 1957 to 1960, Aldon Morris documents the spread of sit-ins as a tactic. He notes how these more confrontational actions were initiated by youth-led organizations that worked in multiple cities during this time. Like the climate movement today, young civil rights activists were frustrated with the lack of results from the more peaceful and nondisruptive activism taking place in the movement. Even though these movements are similar in that young people became frustrated and initiated more confrontational activism, one notable difference is the degree to which the groups engaging in disruptive activism are embedded in and supported by their broader communities.

Like the conflicts we see within the climate movement regarding how to respond to the incremental climate policies coming out of the Biden administration, there were also conflicts among factions within the civil rights movement.[90] Nevertheless, the more confrontational activists in the civil rights movement were still embedded in the broader community of Black Americans. Morris explains how Black churches "supplied these organizations, not only with an established communication network, but also leaders and organized masses, finances, and a safe environment in which to hold political meetings."[91]

In contrast to these direct connections between civil rights activists engaging in civil disobedience and institutions within the Black community, climate groups that are engaging in direct action are working hard to stay anonymous, even from the broader climate movement and other potential progressive allies. In interviews, many shockers told me how they intentionally keep themselves disconnected through an "affinity group structure." When they organize actions, groups "are often working in a black bloc [because they intend to get arrested]. Nobody knows anyone's name, it's all pseudonyms on Signal."[92] One of the activists lamented how this strategy made it much more difficult for creating community and showing solidarity: "How do people know that I will show up for them if everyone's anonymous?"[93]

Because states and countries are applying more aggressive penalties to climate activists in an effort to deter the spread of confrontational tactics,[94] the desire to mask one's identity when engaging in civil disobedience is reasonable. As the penalties increase, so to do activists' efforts to remain anonymous. With the growing risk to activists participating in confrontational tactics, it is hard to imagine how they could safely embed themselves in a community without putting the entire community at risk.

During the civil rights movement, there were also substantial risks to activists participating in the movement.[95] As activists mobilized to challenge repressive policies that kept Black Americans from attaining equal rights, many activists paid a high price for participating in predominantly peaceful activism. Activists were routinely persecuted, and many were met with violent responses from white supremacists and from repressive law enforcement.[96] As counterprotesters and law enforcement responded with violence, the civil rights movement took advantage of the activists' blood in the streets. In fact, McAdam documents how activists in the civil rights movement aimed to provoke "violence as a stimulus to favorable government action."[97] Although the activism during this time was predominantly nonviolent civil disobedience, law enforcement and the countermovement of White Supremacists responded with violence, which was televised across the United States. The violence against peaceful Black activists motivated what some scholars call moral shocks: "when an event or situation raises such a sense of outrage in people that they become inclined toward political action, even in the absence of a network of contacts."[98]

In our conversation about the future of the climate movement, Keya Chatterjee of USCAN spoke of lessons learned from this period in the civil rights movement. "You have to create moments like the Freedom Riders . . . where the middle of the country realizes their values are not aligned with the opposition and they get uncomfortable." Chatterjee specifically discussed the violence against protesters during this period in the civil rights movement: "We are here to change peoples' views . . . The point is they change their views when they see the oppressor overreaching and being so violent."[99]

Although many critiques of confrontational activism focus on public opinion about it, it is important to note that this type of activism is usually disliked by the general public. For example, the Freedom Riders that Chatterjee mentions were quite unpopular during the civil rights movement. Figure 4.1 shows

Freedom Riders, sit-ins (May 1961)

Q: Do you approve or disapprove of what the "Freedom Riders" are doing?

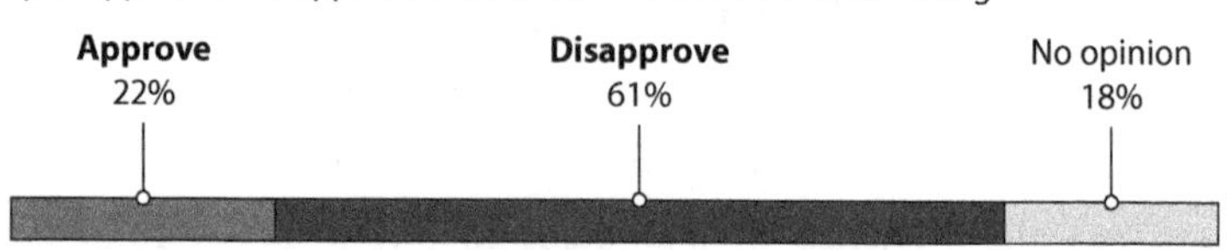

Q: Do you think "sit-ins" at lunch counters, "freedom buses" and other demonstrations by Negroes will hurt or help the Negro's chances of being integrated in the South?

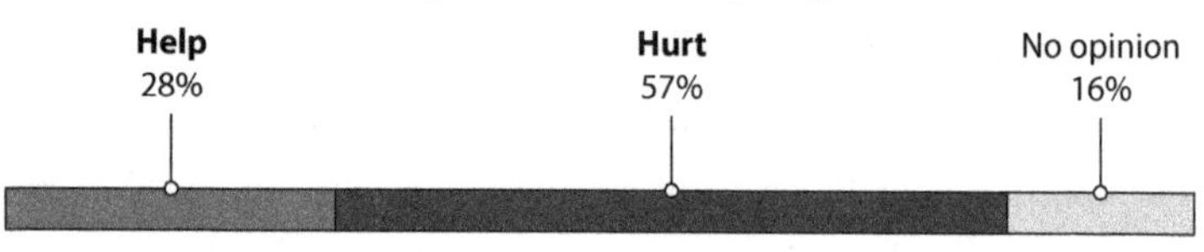

Note: May not equal 100% because of rounding.

FIGURE 4.1 Public opinion in 1961 about the Freedom Riders and sit-ins during the civil rights movement from Gallup data collected May 28–June 2, 1961.

public opinion about the Freedom Riders and the sit-ins based on data collected from Gallup in 1961.

As climate activism gets more confrontational, we should expect that climate activists who employ direct action will also be unpopular. In response to questions about public opinion regarding specific activist tactics, Chatterjee continues: "That's okay. We're not here to be liked." McKibben also acknowledged that civil disobedience is likely to be unpopular with some: "One of the things that you always have to measure when you're doing things like civil disobedience is whether or not it will attract more people or turn more people off."[100]

As the climate crisis worsens and climate shocks grow more frequent and more severe, climate activism and the radical flank will undoubtedly grow. Across the range of groups that I spoke with and their various perspectives on the appropriate level of confrontation, the activists who were engaging in direct action all agreed that much more shock and disruption was needed. As the spokesperson for XR NYC told me: "Civil disobedience needs to be at mass scale in order to work."[101]

Confrontational activism that includes people with a bigger platform, be they actors like Fonda, well-known activists like McKibben and Thunberg, or religious leaders like Rosenn, helps the movement gain more media attention.[102] When these actions are connected to broader sustained campaigns with specific locally embedded targets, their effects can be traced well beyond a wave of media attention. As the climate crisis worsens, the climate movement, including the radical flank of shockers and disruptors, have much to learn from the civil rights period of activism in the United States. Climate campaigns that are more embedded in local communities and broadcast the violence directed against them take advantage of opportunities to connect the issue to local struggles and draw sympathy from the broader public. Until the masses mobilize, however, civil society is limited in what effects it can hope to accomplish.

Chapter Five

SAVING OURSELVES WILL TAKE A DISASTER (OR MANY)

IN JUNE 2023, much of the eastern United States was under an air quality alert. Eighteen states experienced dangerous air as smoke from wildfires burning in Canada blanketed an area from the Great Lakes to the northeastern United States. Famous sights around New York City were shrouded in an orange haze. Further south in the Washington, DC, area, the visibly smokey air had an odd scent. Even though our air didn't look 'Martian' and all outdoor activities had been canceled, my daughter and I both found our asthma flaring whenever we went outside.

Forest fires are not uncommon in Canada. However, the "size, ferocity, and number of fires" burning across the country were attributed to climate change.[1] Canadian prime minister Justin Trudeau took to Twitter (which became X in July 2023) to announce: "We're seeing more and more of these fires because of climate change. These fires are affecting everyday routines, lives and livelihoods, and our air quality."[2] This climate shock affected an estimated 98 million people in the United States, leading to canceled Major League Baseball (MLB) games and indoor recess at schools.[3] The wind blew away the smoke by mid-June, but it returned again-and-again throughout summer 2023. As the fires continued into September, experts predicted that the season would end up being Canada's longest wildfire season ever.[4]

This first-of-its-kind, forest fire–induced air quality crisis hit the northeastern United States less than a week after President Joseph Biden signed a bipartisan deal that resolved the 2023 debt ceiling crisis, which threatened to devastate the American economy.[5] Although it was a big win for the president and more evidence that he could broker an agreement between the Democratic and Republican parties in the leadup to the 2024 election cycle, the compromise dealt a huge loss to the environmental community. The deal rolled back aspects of the National Environmental Protection Act (NEPA) and approved permitting for the Mountain Valley Pipeline (MVP) that would transport natural gas from West Virginia. Beyond expediting permitting for the MVP and insulating the project from judicial review, the deal set a legal precedent for future fossil fuel infrastructure projects.[6]

The debt ceiling agreement put even more strain on groups in the climate movement. While greenlighting the MVP had been a requirement for West Virginia senator Joe Manchin's support of the Inflation Reduction Act (IRA) in summer 2022, the climate community had hoped that the MVP would never be approved. Big and small climate groups publicly admonished the administration as the debt-ceiling deal came together. At the same time, those groups also privately criticized organizations that remained quiet, thus providing implicit support for the president and his energy policy. Once again, the left was infighting and collectively "clapping for Tinkerbell" as the world burned.[7]

Less than two weeks later, tensions escalated further when four Big Green climate groups endorsed President Biden's 2024 bid at a gala in Washington, DC, while other climate activists protested outside.[8] Many groups wondered how an early endorsement gave away any leverage the climate community might have had in the run-up to the election. Others asked how these four Big Green environmental groups could support an administration that ignored the environmental justice concerns of frontline communities and communities of color that it had promised to support.

By the end of June 2023, the list of approved fossil fuel projects belied President Biden's claim that his administration was following through on his campaign pledge to transition away from fossil fuels.[9] The Biden administration's energy strategy put the United States on a path to expand its ability to extract and burn fossil fuels at home and export them abroad for decades to come.[10] The administration's support for continued fossil fuel expansion also made it very difficult to believe that the United States would meet its commitments to the Paris Agreement and do its part to limit climate change.[11]

Most other countries were doing no better.[12] The Group of Seven's "Leaders' Communiqué" from its June 2023 meeting in Hiroshima stressed "the important role that increased deliveries of LNG [liquified natural gas] can play" and noted the group's support for further "investment in the gas sector."[13] At the same time, the *Statistical Review of World Energy* announced that, even with investments in renewables, fossil fuels were holding steady and continuing to supply 82 percent of the world's energy.[14] In the United Kingdom, the independent Climate Change Committee that was formed as part of their Climate Change Act submitted a 2023 progress report that led the *Guardian* to announce that the United Kingdom was "missing climate targets on nearly every front" as it moved forward with new coal, oil, and gas projects.[15] Even Norway, which had publicly applauded itself for successfully transitioning to electric vehicles, extended its fossil fuel production by approving nineteen new oil and gas fields for development.[16]

These recent expansions of fossil fuels offer more evidence that countries are not responding adequately to the call for climate action. Thirty years into attempts to address climate change, vested interests that have amassed wealth and power through the extraction and burning of fossil fuels continue to have a stranglehold on economic and political power. We saw this process play out in the debt ceiling deal in the United States, the debates leading up to the twenty-eighth round of the climate negotiations for the UN Framework Convention on Climate Change Conference of the Parties (COP28), and in other developments around the world. As a result, global carbon dioxide levels continue to rise, along with the rates of methane and nitrous oxide, which the National Oceanic & Atmospheric Administration (NOAA) Global Monitoring Laboratory calls the "three main long-term drivers of climate change."[17]

With so many steps in the wrong direction, it should surprise no one that the world is on track to surpass the 1.5°C threshold codified in the Paris Agreement and built off research from the Intergovernmental Panel on Climate Change (IPCC).[18] Updated analysis that followes the IPCC methodology documents how Earth continues to heat up and experience a "high rate of warming."[19] In September 2023, the World Meteorological Organization (WMO) reported that summer 2023 was Earth's "hottest three-month period on record, with unprecedented sea surface temperatures and much extreme weather."[20] Heat domes shattered temperature records around the world and people experienced climate shocks in the form of floods, droughts, glacial and sea ice melt, marine heat waves, wildfires, and more. The social consequences of these climate shocks were all around us as people tried to adjust to a new normal that included extreme heat and storms while the oceans boiled, the polar ice melted, and Canada and Maui (and many other places around the world) burned.

LOOKING BACK

This book began by presenting an overview of the governmental and business responses to the climate crisis, explaining why progress has been so slow and why the incremental changes we have observed so far are insufficient. I documented how fossil fuel interests have contributed to the sluggish pace of progress. Next, I turned to the role that the people must play in bringing about necessary climate action. I explained how fossil fuel interests have redirected our attention away from the systems that they have captured and then outlined how collective action through activism can push back against these vested interests and respond to the climate crisis. Like the state and market sectors, however, the outcomes of these efforts thus far have not been adequate to meet the challenge. In the last chapter, I provided an overview of the expanding radical flank of the climate movement to understand how activists are employing shock and disruption to try to stop the climate crisis.

This book explains how neither the state, the market, nor the civil society sector has been able to guide society onto a path that meets the international commitments of the 2015 Paris Agreement, even though it was ratified by 195 of the 198 parties to it.[21] The UN secretary general gave a chilling speech about this dire situation in a press conference in June 2023:

> I am very worried about where the world stands on climate. Countries are far off-track in meeting climate promises and commitments . . . The climate agenda is being undermined. At a time when we should be accelerating action, there is backtracking. At a time when we should be filling gaps, those gaps are growing. Meanwhile, the human rights of climate activists are being trampled. The most vulnerable are suffering the most. Current policies are taking the world to a 2.8 degree [Celsius] temperature rise by the end of the century. That spells catastrophe. Yet the collective response remains pitiful. We are hurtling towards disaster, eyes wide open—with far too many willing to bet it all on wishful thinking, unproven technologies and silver bullet solutions.[22]

While nation-states increase penalties for activists engaging in climate protest,[23] current levels of activism are not mobilizing a critical mass that has the potential to pressure the state and/or the market sectors to limit global warming. With these terrifying trends and overwhelming evidence that climate shocks are getting worse, states and businesses are not doing what is needed, and yet collective action has not increased. Instead, there have been fewer protests that engage a smaller number of people since Biden took office. National surveys document that individual concern about the climate crisis has grown,[24] but it has not been enough to motivate action. Although the radical flank has engaged in more civil disobedience—with confrontational actions at public events and against elected officials happening regularly—the number of people participating remains quite limited.

For example, the week after the debt ceiling deal was signed, a climate protest took place outside the White House. Hundreds of activists, including members of Indigenous communities that live along the path of the MVP, older activists in their painted rocking chairs, and the young people who had been getting a lot of attention for shutting down the talks of members of the Biden administration, joined together to chant, sing, and call for the president to stop the MVP and declare a climate emergency.[25] Although the protest included notable activists like Bill McKibben and congressional representative Rashida Tlaib, the underwhelming turnout and the timing of the protest *after* the bill was signed by the president made it hard to imagine how this action would have any effect on the political process.

HOW BAD WILL IT GET?

As I outlined in chapter 1, even though climate shocks have become much more common, social change that is significant enough to address the climate crisis adequately will only come through an AnthroShift initiated once these shocks become even more severe and more frequent. Scientific papers with an escalating tone of urgency trumpet the growing evidence that we are fast approaching abrupt and irreversible climate tipping points.[26] Even as these messages grow louder, countries continue to approve more fossil fuel projects, and companies walk back their climate commitments. *In sum, it's bad, it's getting worse, and nothing we have done comes anywhere close to what is needed.*

Climate shocks will lead to more human suffering. We already see the effects of floods, droughts, fires, and severe weather. As these shocks increase, economies will suffer and those who can move will. Like the experience of illness and death during the COVID-19 pandemic, the suffering from climate shocks will not be felt equally across or within societies; existing inequalities will lead to inequalities in the pain and suffering that will be experienced. The wealthiest may be able to limit their exposure and fly away from the crisis in their private jets, but the climate crisis will grow and eventually everyone everywhere will experience the effects of our warming world.

The COVID-19 pandemic taught us that risk and threat can become normalized over time. People realized that the risk of disease was not universal across society, and the lag from infection to serious illness stalled progress. The same pattern played out in summer 2023. Civic groups and individuals in the northeastern United States responded initially to the forest fire–fueled toxic air with calls to mobilize, but then society became accustomed to cancelling sports practices and staying inside.

As I mentioned in chapter 1, "[r]isk only motivates an AnthroShift if it surpasses a threshold in terms of its duration and intensity. Without a sustained shock that has tangible consequences in terms of social cost to people and property, the subsequent social change (like we saw during the early days of the pandemic) will be ephemeral." Transformational social change requires a massive shock to motivate an AnthroShift that yields the economy-wide changes needed. It is conceivable that the level of shock required to get us to an AnthroShift will involve sections of the world becoming uninhabitable, leading to mass migration as well as pain, suffering, and death around the world. A paper published in *Nature Sustainability* in May 2023 analyzed changes in the "human climate niche," where populations of humans can comfortably survive based on the mean annual temperature. The authors showed that "climate change has already put ~9 percent of people (>600 million) outside this niche," with current policies estimated to leave one-third of the global population living in areas with increased morbidity and mortality.[27]

How much suffering will we need to endure before a critical mass of people rises up to push actively and aggressively for social change? History tells us that the precautionary principle loses when it goes head-to-head against entrenched fossil fuel interests that have benefited from privileged access to power and capital for so long. Rather, it is the universal experience of personal risk that motivates change. As U.S. Deputy Special

Envoy for Climate Jonathan Pershing noted in our conversation: “If people understand the risk, we believe they may be more likely to change the direction of their policy.”[28] Research shows us that it is only after people experience a disaster that windows of opportunities open to implement the type of systemic change that is needed.[29]

As climate shocks come more frequently and with more severity, the associated social upheaval will lead to substantial growth in the climate movement. Experiencing climate shocks firsthand will lead more and more people to decide they can no longer sit back and wait for the state and the market to save them. In the words of 350.org cofounder and climate activist McKibben: “As the climate gets warmer, hotter, and as this chaos just keeps increasing as a result, you're going to see more and more and more of this kind of action.”[30] Another climate leader was more explicit about what he saw as the threshold before a critical mass mobilized: “People really need to feel the immediate loss in terms of whether they have access to food or the price of food and all that. The cost-of-living prices, I think, needs to affect people to really drive them to action.”[31]

These expectations were corroborated by a report by the Deloitte Center for Integrated Research in April 2022, which documents how climate shocks are associated with increased civic engagement.[32] In their analysis of data from twenty-three countries, the authors find that people who had experienced at least one climate event (which they defined as “extreme heat, wildfire, drought, etc.”) over the last six months were much more engaged in personal civic actions around the environment.[33]

HOW DO WE SAVE OURSELVES?

At this point, we cannot stop our warming world, but we can prepare to survive what comes next and work to limit the damage. Many climate activists have become preoccupied by the notion that mobilizing 3.5 percent of the population will bring about the social change they envision. However, as many scholars have noted (including the original authors of the work that introduced this famous metric), the statistic is based on historical observations where it did not always hold.[34] It is also unclear if that level of mass mobilization will lead to the desired policy changes in advanced industrialized nations. We have yet to see this level of mobilization in most countries—in the United States, for example, 3.5 percent of the population is over 11.5 million people. It is hard to imagine that this percentage of the United States, or other countries, would mobilize to engage in climate activism without some sort of large-scale disaster as motivation.

What will this activism look like? Will it be confrontational, involving shock and disruption? Only time will tell. There is reason to believe, however, that a mass mobilization of civil society would accelerate the pace of the necessary changes to bring about an AnthroShift and limit human suffering associated with the climate crisis.

History shows us how civil society can take back power and pressure decision makers to change policies when its members feel they have no other choice. A sustained mass mobilization, even if it is nonviolent, can motivate political change because decision makers are threatened with a legitimation crisis.[35]

Social movements have that power, but they will need a critical mass that pressures both the market sector and the government to accomplish the level of systemic changes that are needed.

Based on the findings presented in this book, which draw from my research on climate action and climate activism, as well as activism and engagement more generally, I have three clear recommendations about what we can do to build our capacity to save ourselves: we should focus our collective efforts on: (1) creating community, (2) capitalizing on moral shocks, and (3) cultivating resilience. The remainder of this book provides details.

CREATE COMMUNITY (AND REAL SOLIDARITY)

Embedding climate activism in community rather than yelling at strangers is a good start. As I have written about in many places over the years, a lot of evidence shows that reaching out to convince strangers to care about and act on your cause by knocking on their doors, calling them, or even sending them a text message *might* get them to turnout for an election. However, parachuting into places where you have no personal ties does not create any lasting connections and certainly does not build community.[36] It is well documented that calling up a random person from a list during dinner to ask them for something (money, support, a vote) does not create the type of dense connections that strengthens democracy. Instead, building on preexisting ties among friends, neighbors, fellow students, or even acquaintances has the potential to create more impact and last longer.

The climate movement should learn from the experiences of the civil rights movement. As I noted in chapter 4, the civil rights movement built its networks through and among Black churches, which were already a trusted and established institution within Black communities around the United States. The churches connected concerned individuals to a range of groups involved in the struggle for racial justice. To repeat the important finding from Aldon Morris's research, it "supplied these organizations not only with an established communication network, but also leaders and organized masses, finances, and a safe environment in which to hold political meetings."[37]

The climate movement should also work to build networks of climate-concerned individuals that are embedded in the communities where they live, work, and experience climate shocks. So much more is possible when people already know one another and have reciprocal social ties.[38] Although the climate movement does not have a specific church or other religious institution, environmental stewardship is a priority for various religious denominations. In recent years, in fact, the Church of England and the Catholic Church have both publicly announced their support for a transition away from fossil fuels.[39] Like other left-leaning groups and movements, however, the climate movement will have to overcome the persistent lack of social infrastructure on the left. As I noted in *Activism, Inc.* in 2006 and repeated in *American Resistance* in 2019, the left and the Democratic Party has continued to lean into laying sod while Republicans cultivate grassroots.[40]

As climate shocks get stronger, there is no question that communities with strong social capital will be better prepared to adapt and recover. As research on natural disasters shows, "communal resources, cohesion, and social infrastructure can mitigate shocks and enhance resilience."[41]

CAPITALIZE ON MORAL SHOCKS (INCLUDING VIOLENCE)

Chapter 4 also discussed how violence against activists in the civil rights movement induced moral shocks among the broader population that helped the movement mobilize support and achieve some of its goals.[42] More recently moral shocks helped to motivate solidarity for the Black Lives Matter movement after unarmed George Floyd was murdered by a Minneapolis, Minnesota, police officer in May 2020. The protests that broke out around the United States in summer 2020 have been called the broadest sustained protests in U.S. history.[43]

There are many reasons why the protests in 2020 were different from previous waves of the Black Lives Matter movement.[44] One of the most notable is that, this time (in contrast to previous instances where police killed an unarmed Black person), the United States witnessed the murder of Floyd through bystander videos that were shared on social media. There is no question that seeing an unarmed Black man die while begging for his life was a moral shock for many who watched. The moral shock was so strong that it motivated numerous political and social movement organizations that were focused on other issues to call for their members to join the protests in solidarity with the Black Lives Matter movement. Many individuals with no connections to social movement organizations or civic groups also turned out to participate.

In many ways, the mass mobilization that we saw in summer 2020 followed the same model that Keya Chatterjee from U.S. Climate Action Network (USCAN) referred to when she described what the climate movement should learn from the civil rights movement: "they change their views when they see the oppressor overreaching and being so violent."[45] Witnessing violence can motivate sympathizers who care about an issue but are not connected through social movement organizations and embedded in specific affected communities.[46]

Even though the long-term effects of the Black Lives Matter movement after the murder of Floyd have been questioned,[47] the climate movement would be wise to learn from this moment in the struggle against systemic racism in the United States. By mobilizing such large crowds in so many locations across the United States, the movement had the potential to motivate greater social change. These collective efforts, coupled with identity-based motivations and the moral shock of witnessing the murder of an unarmed Black man by a police officer over social media, provided a dynamic catalyst for participation across race, gender, sexual orientation, and other identities.[48]

The climate movement should also work harder to connect with the growing labor movement. In recent years, the United States has experienced a labor renaissance, with workers at Amazon, Starbucks, newspapers, and even university graduate students unionizing.[49] While the overall numbers of unionized labor continue to be limited,[50] labor is flexing its muscles in response to growing inequality. As the world warms and climate shocks get worse, inequality will continue to increase, and labor will become more aligned with the plight to save ourselves. The climate movement would benefit from putting the crisis in a context that resonates with workers.

While numerous climate leaders mentioned to me that labor needed to be more engaged in the struggle to stop the climate crisis, others noted the challenges. For example, the executive director from the BlueGreen Alliance, which is a coalition of labor unions and environmental groups, spoke about how some climate activism put off workers and organized labor: "fights focused on fossil fuel infrastructure have been absolutely divisive and toxic for the efforts to unite the labor and environmental movements."[51]

Recent research documents how labor is not just threatened by activism against fossil fuel infrastructure. In their study of the electric vehicle industry, Jennifer M. Silva, Sanya Carley, and David M. Konisky find that "[w]orkers suspect that their past contributions will be forgotten and express suspicion about the green energy transition."[52] This research underscores the continued need to build bridges across identities, orientations, and occupations to develop support for an energy transition.

CULTIVATE RESILIENCE

Because mitigation efforts have so far failed to stop the climate crisis, we must all be prepared for what is coming.[53] Saving ourselves requires that we adapt socially and environmentally to a world of more frequent and more severe climate shocks. *Everyone*, not just activists, should be prepared for what is coming so we can create resilience in our communities and in the environments around them.

A growing number of service corps programs are being coordinated around the country with the mission to train people to work on climate resilience in their communities for pay. These programs are expanding at the federal level and popping up at the state and even local levels all over the United States. The work takes many forms, including planting trees in communities to reduce the heat island effect,[54] limiting stormwater runoff,[55] and absorbing carbon, as well as helping rebuild communities after natural disasters. The projects have the potential to motivate local climate action that prepares individuals and communities to adapt to climate change and support local populations as they feel the effects of climate shocks.

As the number of these programs grow, they must effectively educate about the climate crisis so participants understand what they are doing and why they are doing it. Research shows that participating in bottom-up, citizen-led adaptation initiatives is associated with an increased awareness about and level of preparedness for climate change impacts.[56]

The agencies and institutions that run these projects must understand the ways that climate mitigation and adaptation are related and how social and environmental resilience can be cultivated and maintained. When done right, these programs can play a vital role in preparing the public to adapt to and be resilient when climate shocks come. In other words, the programs have the potential to play an important role in preparing us to save ourselves—and we are going to need all the help we can get.

There is no question that, as climate shocks get more severe and come more frequently, more people will rise up in response. If they can connect with community, capitalize on moral shocks to create solidarity, and cultivate resilience, there is no doubt that they will be able to harness their capacity more quickly to encourage and support an AnthroShift that leads to the necessary systemic changes. I believe that the more coordinated these efforts are, the less people will suffer and die before we get to the other side.

As unfair as it might seem, the future is up to us.

METHODOLOGICAL APPENDIX

THIS PROJECT utilized a multistage and mixed methodological approach to understanding progress by various social actors to *save ourselves*. It synthesizes data collected over the past twenty-five years. In places where I discuss findings that were already published in peer-reviewed publications, I specifically cite the original publication, which provides details on the data and research methods used. In addition, data were collected through surveys and interviews with members of the U.S. climate policy network in 2022, surveys with activists and organizers around the 2023 Earth Day actions, and open-ended semi-structured interviews with various groups and organizers in 2022 and 2023 about the growing radical flank in the climate movement. This appendix provides a detailed description of the new data and analysis that are included in this book.

STUDYING THE EVOLVING CLIMATE POLICY NETWORK

CONSTRUCTING A DATA SET AND SAMPLING

Consistent with previous research studying echo chambers in climate policy networks, sampling for this project began by creating a comprehensive data set of elite U.S. climate policy actors engaged in policymaking during the period of the study.[1] A data set was assembled from three publicly available sources. My research team and I began with a list of all the policy actors who participated in climate-related hearings in the U.S. Congress during the two sessions prior to when data collection began: the 115th session (January 2017 to January 2019) and the 116th session (January 2019 to January 2021). Consistent with previous studies using this data source,[2] a search for all hearings that discussed climate change was conducted through the Government Printing Office (GPO) FDSys search engine, which archives transcripts from congressional hearings and makes them available for the public record. Using the search terms "global warming," "greenhouse gas," and "climate change," we identified all the hearings that discussed these issues during the 115th and 116th sessions of the U.S. Congress (2017–2021).

The searches yielded a total of 1,065 hearings that fit the search term criteria: 512 hearings from the 115th session of the U.S. Congress and 553 from the 116th session of the U.S. Congress. The contents of each hearing were then reviewed to confirm that the focus of each hearing was actually the topic of climate change. Of the 1,065 hearings that mentioned the search terms, eight-two hearings were deemed to have the relevant focus on climate change. From these hearings, 146 testimonies were delivered in eighteen separate hearings in the 115th session of Congress, and 516 testimonies were delivered in sixty-four separate hearings in the 116th session of the U.S. Congress. Only formal statements were included in the analysis. Comments made during the question-and-answer portion of the hearings were not analyzed.

Next, we tabulated all nongovernment actors who were registered in the Senate and House to lobby on climate issues during each respective period through the Open Secrets Lobbying database.[3] We counted any organization that made payments between 2017 and 2020.[4] Finally, we cross-referenced this list with a roster of all attendees from the United States who participated in the international climate change negotiations (the twenty-first UN Climate Change Conference of the Parties [COP21]) in Paris in December 2015.[5] By drawing from these varied sources, we were able to assemble a data set that measured sustained engagement in the U.S. climate policy network over the years leading up to our period of data collection. It would have been ideal to include a list of the speakers who were participating in the ongoing 117th session of the U.S. Congress as well, but data collection was taking place in the middle of this particular session, and thus a complete list of hearings and participants was not available.

Actors in this data set were ranked according to the degree to which they participated in hearings, international negotiations, and lobbyist registries (if they were nonstate actors). Testimonies were weighted so that multiple appearances before

Congress indicated greater participation. In some cases, policy actors participated in climate-related congressional hearings more than once, but in other cases, the actors participated in a congressional hearing once but also participated in the climate negotiations and/or were registered to lobby on the issue. After reviewing the rankings of actors across the four-year period, we were unable to identify a defensible threshold for participating in the policy network. When we included the 115th session, the emergent thresholds either yielded a sample that was too large (155 actors) or too small (59 actors). As a result, we decided to remove participation in the 115th session from the sampling criteria. After taking that additional step, a clear threshold was identifiable that was relatively consistent with previous samples of policy actors who had been actively involved in climate-related discussions during this period of time. The 116th session of the U.S. Congress was the first session to include the Select Committee on the Climate Crisis in the House of Representatives so there were ample climate-related discussions during this period. Thus, our final sample for this wave of data collection includes all 110 actors who participated more than once in this "climate policy arena."

Policy actors fell into eight *types*: (1) businesses and business associations/trade groups; (2) Democratic members of the U.S. Congress; (3) environmental groups; (4) nongovernmental organizations (NGOs), including professional associations and think tanks; (5) Republican members of the U.S. Congress; (6) scientists; (7) subnational governmental representatives; and (8) U.S. executive branch, including representatives from government agencies.

DATA COLLECTED

The data analyzed for this book were collected between February and April 2022. In my analysis, I compare these data to the data collected during previous waves of this project in 2010, 2016, and 2017. It is worth noting that the 2010 data collection period was a period when climate policy was working its way through the U.S. Congress (and was eventually unsuccessful), the 2016 data collection period involved debate about President Barack Obama's Clean Power Plan, and the 2017 data collection period took place while the Trump administration was dismantling components of the Obama administration's Climate Action Plan and withdrawing the United States from the Paris Agreement.[6] The 2022 wave of data collection took place while the United States Congress was considering the Build Back Better Act, which was passed in a substantially reduced form as the Inflation Reduction Act (IRA) in August 2022.[7]

Data collection was conducted in accordance with University of Maryland (UMD) policies on human subjects research (institutional review board [IRB] Protocol number 1838270-1). The 110 policy actors identified in the sample were contacted to participate in the study because they represent the core of political elites who have the most influence over the policy process. Because the COVID-19 pandemic was ongoing, data were collected through meetings over Zoom. Actors were interviewed and administered a survey on Qualtrics. Both the survey and interview data inform the findings in this book. In total, survey data were collected from seventy policy actors, and interview data were collected from sixty-eight policy actors in our sample, representing a

TABLE A.1 Survey and interview data from 2022

Organizational Type	Surveys, N (%)	Interviews, N (%)
Businesses and business associations/trade groups	20 (56%)	18 (50%)
Congressional Republicans	8 (53%)	7(47%)
Congressional Democrats	5 (56%)	5 (56%)
Environmental groups	13 (81%)	13 (81%)
Nongovernmental organizations (NGOs)	13 (87%)	13 (87%)
U.S. executive branch, including government agencies	3 (100%)	3 (100%)
Subnational governmental representatives	6 (43%)	7 (50%)
Scientists	2 (100%)	2 (50%)
Total	**70**	**68**

64 percent and 62 percent response rate, respectively. These response rates are consistent with previous waves of the study.[8]

Even though respondents included actors from across the political spectrum and from all types of organizational affiliations, I observed some differences between the respondents and the nonrespondents in the study. Table A.1 includes the distribution of respondents by organizational type.

SURVEY INSTRUMENT

The survey itself was comprised of attitudinal questions and network questions. Attitudinal questions asked participants to indicate, on a scale of 1 to 5, where 1 indicated strong disagreement, 5 indicated strong agreement, and 3 indicated neutrality, their positions on statements that were deemed politically salient during the data collection period. For the network question, each of the 110 policy actors in the sample was listed in alphabetical order by actor type, and each respondent was presented with three iterations of this list. Participants were then asked to indicate, in order, those actors whom they identified as their sources of expert scientific information about climate change; those actors or organizations they collaborate with on a regular basis; and those actors whom they perceived to be most influential in climate politics, in any ideological direction.

POLICY ACTOR INTERVIEWS

All policy actors were also asked to participate in a short interview. Interviews followed an open-ended, semi-structured format and were more conversational than scripted.[9] Interviews followed a predetermined protocol, which focused on the status of climate and energy policy in the United States. Each policy actor was asked to describe what their organization or office considered the ideal climate policy and then to discuss the tactics it determined would be the most effective for achieving their goals. This interview method allowed for flexibility, encouraging the interviewer to pursue follow-up questions and the interviewee to express candid responses. Interviews ranged from twenty to ninety minutes; they were recorded digitally and transcribed prior to analysis. Transcripts of the interviews were coded by hand into broad themes and then recoded into what were determined to be relevant subthemes.

STUDYING THE EXPANDING CLIMATE MOVEMENT

INTERVIEWS WITH ACTIVISTS AND ORGANIZERS

In addition to these extensive data on the U.S. climate policy network, which represent groups that have been working specifically to affect federal climate policymaking in the United States as insiders to the institutional political process, I also interviewed activists and organizers working as outsiders in the policymaking process, I conducted interviews with organizers and activists from the growing radical flank of the climate movement in summer 2022 and 2023. Groups were identified through a snowball sample that was drawn from my interviews with policy actors in 2022, through cosponsorship of particular protests, or media mentions. These interview data followed the same protocol as the interviews with the climate policy network. Data were collected in accordance with the University of Maryland's policies on research on human subjects (UMD IRB Protocol number 1838270-1).

Interviews included the same questions asked of NGOs who were included as part of the climate policy network analysis. Specifically, activists and organizers were asked to reflect on their perspective on the ideal U.S. climate policy, as well as what tactics were necessary to achieve those policy outcomes. They were also asked about their personal activist backgrounds and the history of the groups in which they were working. In addition, they were asked to describe the range of tactics they used in their activism, how and why they chose these tactics, and what were the intended goals of the activism. Interviews lasted from thirty to ninety minutes.

In all cases, interviews were open-ended and semi-structured. The purpose of the semi-structured interviewing technique, as summarized by John Lofland and Lyn H. Lofland, is "to achieve analyses that (1) are attuned to aspects of human group life, (2) depict aspects of that life, and (3) provide perspectives on that life that are simply not available to or prompted by other methods of research."[10] The interviews were recorded and transcribed, and extensive notes and memos from all the interviews were kept as the bulk of the qualitative data set. As the patterns across cases emerged, I distinguished between first-order conclusions (i.e., those explicitly drawn or stated by the respondent), and second-order conclusions (i.e., those drawn from what was said). In so doing, I acknowledge my role in interpreting the data patterns, as well as subjecting the respondents' claims to additional scrutiny.

When I am referring to interviews with representatives of groups who agreed to speak on the record, I reference the name of the person, his or her affiliation, and the date of the interview. For those people who spoke with me with the understanding that they would not be directly attributed or for those quotes that speakers did not want directly referenced to them, I reference those conversations without naming the individual or their group. Quotations included in this book have been edited to remove repetitive phrases and words such as "y'know," "like," "um," and "kind of," which have no bearing on the content of the statements. In cases where these words provide additional meaning to the quotes, they were not deleted.

SURVEYS WITH CLIMATE ACTIVISTS AT THE 2023 EARTH DAY ACTION

In addition to the data collected at the two People's Climate Marches (PCMs) and the youth-led climate strikes in 2019 and 2020 (the methodology for which are described in papers referenced in the book), data were also collected at the 2023 Earth Day Action in Washington, DC. Because the action was relatively small, my five-person research team aimed to sample as many people in the crowd as possible—and specifically targeted at least one person from each group standing on Freedom Plaza before and during the rally that day. Members of the five-person research team were placed in the crowd at incremental positions to gather data throughout the area. This method avoids the potential of selection bias by preventing researchers from selecting only "approachable peers."[11] It was slightly different from the method that I have used at larger protests. It is worth mentioning that there was a group of activists at the action who did not speak English. Because the survey instrument was not translated to Spanish, we were unable to collect a representative sample from that group. Beyond that, however, we appeared to hit saturation. By the end of the rally, most participants sampled reported having been contacted by one of the researchers from my research team who was surveying in the crowd.

Researchers completed surveys with 143 participants, representing a response rate of 79 percent. Although this refusal rate is higher than at many protests where I have collected data, it is mostly because we did not have a survey available in Spanish and because some people who had attended to engage in civil disobedience were uncomfortable filling out a survey.

The survey was designed to be short and noninvasive to encourage the highest level of participation possible and to facilitate data collection in the field: it took about ten minutes for participants to complete it. Similar to recent studies,[12] data were collected via electronic tablets and quick-response (QR) codes that connected respondents to an online Qualtrics survey at the University of Maryland. These survey data were collected in accordance with the University of Maryland policies instituted by their institutional review board (UMD IRB Protocol number 1439617-8). Individuals age twelve and older were eligible to participate in the study.

NOTES

1. NO ONE ELSE IS GOING TO SAVE US: UNDERSTANDING THE SOCIAL SIDE OF THE CLIMATE CRISIS

1. D. Kriebel et al., "The Precautionary Principle in Environmental Science," *Environmental Health Perspectives* 109, no. 9 (September 2001): 871, https://doi.org/10.1289/ehp.01109871; see also Kenneth R. Foster, Paolo Vecchia, and Michael H. Repacholi, "Science and the Precautionary Principle," *Science* 288, no. 5468 (May 12, 2000): 979–81, https://doi.org/10.1126/science.288.5468.979.
2. Charlie J. Gardner and Claire F. R. Wordley, "Scientists Must Act on Our Own Warnings to Humanity," *Nature Ecology & Evolution* 3, no. 9 (September 2019): 1271–72, https://doi.org/10.1038/s41559-019-0979-y; Charlie J. Gardner et al., "From Publications to Public Actions: The Role of Universities in Facilitating Academic Advocacy and Activism in the Climate and Ecological Emergency," *Frontiers in Sustainability* 2 (2021), https://www.frontiersin.org/articles/10.3389/frsus.2021.679019; Stuart Capstick et al., "Civil Disobedience by Scientists Helps Press for Urgent Climate Action," *Nature Climate Change* (August 29, 2022): 1–2, https://doi.org/10.1038/s41558-022-01461-y; Fernando Racimo et al., "The Biospheric Emergency Calls for Scientists to Change Tactics," ed. Peter Rodgers, *ELife* 11 (November 7, 2022): e83292, https://doi.org/10.7554/eLife.83292.

3. Earth Observatory, "World of Change: Global Temperatures," accessed September 1, 2023, https://earthobservatory.nasa.gov/world-of-change/global-temperatures.
4. For a discussion of climate shocks, see Idean Salehyan and Cullen S. Hendrix, "Climate Shocks and Political Violence," *Global Environmental Change* 28 (September 1, 2014): 239–50, https://doi.org/10.1016/j.gloenvcha.2014.07.007; Meredith T. Niles and Jonathan D. Salerno, "A Cross-Country Analysis of Climate Shocks and Smallholder Food Insecurity," *PLOS ONE* 13, no. 2 (February 23, 2018): e0192928, https://doi.org/10.1371/journal.pone.0192928; Samuel Sellers and Clark Gray, "Climate Shocks Constrain Human Fertility in Indonesia," *World Development* 117 (May 1, 2019): 357–69, https://doi.org/10.1016/j.worlddev.2019.02.003; see also Gernot Wagner and Martin L. Weitzman, *Climate Shock: The Economic Consequences of a Hotter Planet* (Princeton, NJ: Princeton University Press, 2015).
5. Full transcript available at United Nations Secretary-General, "UN Secretary-General's Press Stakeout with Prime Minister Sharif at the International Conference on a Climate Resilient Pakistan," January 9, 2023, https://www.un.org/sg/en/content/sg/press-encounter/2023-01-09/un-secretary-generals-press-stakeout-prime-minister-sharif-the-international-conference-climate-resilient-pakistan.
6. Sarah Kaplan and Andrew Ba Tran, "Nearly 1 in 3 Americans Experienced a Weather Disaster This Summer," *Washington Post*, September 4, 2021, https://www.washingtonpost.com/climate-environment/2021/09/04/climate-disaster-hurricane-ida/.
7. Jeff Beradelli, "Climate Classroom: Is Climate Change Impacting Hurricanes?," October 8, 2022, https://www.wfla.com/weather/climate-classroom/climate-classroom-is-climate-change-impacting-hurricanes/.
8. "Biggest Climate Toll in Year of 'Devastating' Disasters Revealed," *Guardian*, December 27, 2022, sec. Environment, https://www.theguardian.com/environment/2022/dec/27/biggest-climate-toll-in-year-of-devastating-disasters-revealed.
9. Christopher Flavelle, Jill Cowan, and Ivan Penn, "Climate Shocks Are Making Parts of America Uninsurable: It Just Got Worse," *New York Times*, May 31, 2023, sec. Climate, https://www.nytimes.com/2023/05/31/climate/climate-change-insurance-wildfires-california.html.
10. Scott Dance, "El Niño Is Looming: Here's What That Means for Weather and the World.," *Washington Post*, May 1, 2023, https://www

.washingtonpost.com/weather/2023/05/01/el-nino-la-nina-weather-climate/.

11. World Meteorological Organizations, "Global Temperatures Set to Reach New Records in Nest Five Years," May 17, 2023, https://public.wmo.int/en/media/press-release/global-temperatures-set-reach-new-records-next-five-years. For an overview of the report and what it means, see Nicola Jones, "When Will Global Warming Actually Hit the Landmark 1.5°C Limit?," *Nature* (May 19, 2023), https://doi.org/10.1038/d41586-023-01702-w.
12. Andrew Curry, "How the Ukraine War Is Accelerating Germany's Renewable Energy Transition," *National Geographic*, May 6, 2022, https://www.nationalgeographic.com/environment/article/how-the-ukraine-war-is-accelerating-germanys-renewable-energy-transition.
13. Vera Eckert and Tom Sims, "Energy Crisis Fuels Coal Comeback in Germany," Reuters, December 16, 2022, https://www.reuters.com/markets/commodities/energy-crisis-fuels-coal-comeback-germany-2022-12-16/.
14. Frank Jordans, "Police Start Clearing German Village Condemned for Coal Mine," AP News, January 11, 2023, https://apnews.com/article/climate-and-environment-germany-europe-business-8cacc38aac7b336a17e3044b15675b3e; see also Loveday Morris, "Germany Portrays Itself as a Climate Leader: But It's Still Razing Villages for Coal Mines," *Washington Post*, October 23, 2021, https://www.washingtonpost.com/world/2021/10/23/germany-coal-climate-cop26/.
15. See, for example, OCHA IFRC and Climate Centre, "Extreme Heat: Preparing for Heatwaves of the Future," October 2022, https://www.ifrc.org/document/extreme-heat-preparing-heat-waves-future; Intergovernmental Panel on Climate Change (IPCC) Working Group 2, *Climate Change 2022: Impacts, Adaptation, and Vulnerability: Contribution of Working Group II to the Sixth Assessment Report of the Intergovernmental Panel on Climate Change* (Cambridge: Cambridge University Press, 2022), https://www.ipcc.ch/report/ar6/wg2/; Matthew Rodell and Bailing Li, "Changing Intensity of Hydroclimatic Extreme Events Revealed by GRACE and GRACE-FO," *Nature Water*, March 13, 2023, 1–8, https://doi.org/10.1038/s44221-023-00040-5.
16. See, for example, Shakeel Ahmad Bhat et al., "Impact of COVID-Related Lockdowns on Environmental and Climate Change Scenarios," *Environmental Research* 195 (April 1, 2021): 110839, https://doi.org/10.1016/j.envres.2021.110839; Fenzhen Su et al., "Rapid Greening Response of

China's 2020 Spring Vegetation to COVID-19 Restrictions: Implications for Climate Change," *Science Advances* 7, no. 35 (2021): eabe8044, https://doi.org/10.1126/sciadv.abe8044; Zander S. Venter et al., "COVID-19 Lockdowns Cause Global Air Pollution Declines," *Proceedings of the National Academy of Sciences* 117, no. 32 (August 11, 2020): 18984–90, https://doi.org/10.1073/pnas.2006853117.

17. For a full discussion, see Benjamin K. Sovacool, Dylan Furszyfer Del Rio, and Steve Griffiths, "Contextualizing the Covid-19 Pandemic for a Carbon-Constrained World: Insights for Sustainability Transitions, Energy Justice, and Research Methodology," *Energy Research & Social Science* 68 (October 1, 2020): 101701, https://doi.org/10.1016/j.erss.2020.101701. See Seyed Ehsan Hosseini, "An Outlook on the Global Development of Renewable and Sustainable Energy at the Time of COVID-19," *Energy Research & Social Science* 68 (October 1, 2020): 101633, https://doi.org/10.1016/j.erss.2020.101633 for an assessment of the ways the pandemic affected the transition to sustainable energy.
18. As quoted in Laurie Goering, "Greta Thunberg Says Coronavirus Shows World Can Act Fast on Crises," Global Citizen, March 25, 2020, https://www.globalcitizen.org/en/content/climate-activist-greta-thunberg-coronavirus/.
19. Dana R. Fisher and Andrew K. Jorgenson, "Ending the Stalemate: Toward a Theory of Anthro-Shift," *Sociological Theory* 37, no. 4 (December 1, 2019): 342–62, https://doi.org/10.1177/0735275119888247; see also Dana R. Fisher, "AnthroShift in a Warming World," *Climate Action* 1, no. 9 (May 9, 2022), https://doi.org/10.1007/s44168-022-00011-8.
20. See, for example, Ulrich Beck, *World Risk Society* (Malden, MA: Polity, 1999); Ulrich Beck, Anthony Giddens, and Scott Lash, *Reflexive Modernization: Politics, Tradition and Aesthetics in the Modern Social Order* (Stanford, CA: Stanford University Press, 1994); Ulrich Beck, "Subpolitics: Ecology and the Disintegration of Institutional Power," *Organization & Environment* 10, no. 1 (March 1, 1997): 52–65, https://doi.org/10.1177/0921810697101008. See also Dean Curran, "Risk Society and the Distribution of Bads: Theorizing Class in the Risk Society," *British Journal of Sociology* 64, no. 1 (March 1, 2013): 44–62, https://doi.org/10.1111/1468-4446.12004; James R. Elliott and Scott Frickel, "Urbanization as Socioenvironmental Succession: The Case of Hazardous Industrial Site Accumulation," *American Journal of Sociology* 120, no. 6 (May 2015): 1736–77; Merryn Ekberg, "The Parameters of the Risk

Society: A Review and Exploration," *Current Sociology* 55, no. 3 (May 1, 2007): 343–66, https://doi.org/10.1177/0011392107076080; Eugene Rosa, Aaron McCright, and Ortwin Renn, *The Risk Society Revisited: Social Theory and Risk Governance*, repr. ed. (Philadelphia: Temple University Press, 2015). For a full discussion regarding how the AnthroShift is related to reflexive modernization, see Fisher and Jorgenson, "Ending the Stalemate."

21. Ekberg, "The Parameters of the Risk Society," 343. See also Robert A. Stallings, "Media Discourse and the Social Construction of Risk," *Social Problems* 37, no. 1 (1990): 80–95, https://doi.org/10.2307/800796; Thomas A. Birkland, *After Disaster: Agenda Setting, Public Policy, and Focusing Events* (Washington, DC: Georgetown University Press, 1997); Arun Agrawal, "A Positive Side of Disaster," *Nature* 473, no. 7347 (May 2011): 291–92, https://doi.org/10.1038/473291a.
22. See particularly Nathan E. Hultman, David M. Hassenzahl, and Steve Rayner, "Climate Risk," *Annual Review of Environment and Resources* 35, no. 1 (2010): 283–303, https://doi.org/10.1146/annurev.environ.051308.084029. For an overview of risk communication, see Dominic Balog-Way, Katherine McComas, and John Besley, "The Evolving Field of Risk Communication," *Risk Analysis* 40, no. S1 (2020): 2240–62, https://doi.org/10.1111/risa.13615.
23. Fisher interview with Jonathan Pershing, February 11, 2022.
24. David John Frank, Ann Hironaka, and Evan Schofer, "The Nation-State and the Natural Environment over the Twentieth Century," *American Sociological Review* 65, no. 1 (2000): 96–116, https://doi.org/10.2307/2657291; Frederick H. Buttel, "World Society, the Nation-State, and Environmental Protection: Comment on Frank, Hironaka, and Schofer," *American Sociological Review* 65, no. 1 (2000): 117–21, https://doi.org/10.2307/2657292; Arthur P. J. Mol and Fredrick H. Buttel, "The Environmental State Under Pressure: An Introduction," in *The Environmental State Under Pressure*, ed. P. J. Mol Arthur and H. Buttel Frederick, vol. 10, *Research in Social Problems and Public Policy* (Bingley, UK: Emerald Group, 2002), 1–11, https://doi.org/10.1016/S0196-1152(02)80003-5.
25. Fisher and Jorgenson, "Ending the Stalemate," 250.
26. Zhu Liu et al., "Monitoring Global Carbon Emissions in 2021," *Nature Reviews Earth & Environment*, March 21, 2022, 1–3, https://doi.org/10.1038/s43017-022-00285-w.
27. Liu et al., 1.

28. Intergovernmental Panel on Climate Change Working Group 2, *Climate Change 2022: Impacts, Adaptation, and Vulnerability.*
29. Intergovernmental Panel on Climate Change Working Group 2, SPM-35.
30. See also Rubén D. Manzanedo and Peter Manning, "COVID-19: Lessons for the Climate Change Emergency," *Science of The Total Environment* 742 (November 10, 2020): 140563, https://doi.org/10.1016/j.scitotenv.2020.140563.
31. Agrawal, "A Positive Side of Disaster;" Kendra McSweeney and Oliver T. Coomes, "Climate-Related Disaster Opens a Window of Opportunity for Rural Poor in Northeastern Honduras," *Proceedings of the National Academy of Sciences* 108, no. 13 (March 29, 2011): 5203–8, https://doi.org/10.1073/pnas.1014123108. See also Thomas A. Birkland, *Lessons of Disaster: Policy Change after Catastrophic Events* (Washington, DC: Georgetown University Press, 2006).
32. See especially Intergovernmental Panel on Climate Change (IPCC) Working Group 3, *Climate Change 2022: Mitigation of Climate Change. Contribution of Working Group III to the Sixth Assessment Report of the Intergovernmental Panel on Climate Change* (Cambridge: Cambridge University Press, 2022), doi: 10.1017/9781009157926. For a broad overview of the pathways and options for mitigation and adaptation, see also Andrew K. Jorgenson et al., "Social Science Perspectives on Drivers of and Responses to Global Climate Change," *WIREs Climate Change* 10, no. 1 (2019): e554, https://doi.org/10.1002/wcc.554.
33. Isak Stoddard et al., "Three Decades of Climate Mitigation: Why Haven't We Bent the Global Emissions Curve?," *Annual Review of Environment and Resources* 46, no. 1 (2021): 653–89; Fergus Green and Harro van Asselt, "COP27 Flinched on Phasing out 'All Fossil Fuels': What's Next for the Fight to Keep Them in the Ground?," The Conversation, November 22, 2022, http://theconversation.com/cop27-flinched-on-phasing-out-all-fossil-fuels-whats-next-for-the-fight-to-keep-them-in-the-ground-194941; Gokul Iyer et al., "Ratcheting of Climate Pledges Needed to Limit Peak Global Warming," *Nature Climate Change* 12, no. 12 (December 2022): 1129–35, https://doi.org/10.1038/s41558-022-01508-0.
34. Andrew Jordan et al., "The Political Challenges of Deep Decarbonisation: Towards a More Integrated Agenda," *Climate Action* 1, no. 1 (March 18, 2022): 6, https://doi.org/10.1007/s44168-022-00004-7. See also Intergovernmental Panel on Climate Change Working Group 3, *Climate Change 2022: Mitigation of Climate Change.*

35. United Nations Environment Programme (UNEP), *Emissions Gap Report 2021* (Nairobi, Kenya: UNEP, October 25, 2021), http://www.unep.org/resources/emissions-gap-report-2021.
36. Next Climate Institute, "Despite Glasgow Climate Pact 2030 Climate Target Updates Have Stalled," June 2022, https://climateactiontracker.org/publications/despite-glasgow-climate-pact-2030-climate-target-updates-have-stalled/. See also Jonas M. Nahm, Scot M. Miller, and Johannes Urpelainen, "G20's US$14-Trillion Economic Stimulus Reneges on Emissions Pledges," *Nature* 603, no. 7899 (March 2022): 28–31, https://doi.org/10.1038/d41586-022-00540-6.
37. UN News, "Guterres Announces 'No Nonsense' Climate Action Summit; Calls for Practical Solutions," December 19, 2022, https://news.un.org/en/story/2022/12/1131842. See also Michelle Nichols, "U.N. Chief to Convene 'No-Nonsense' Climate Summit in 2023," Reuters, December 19, 2023, https://www.reuters.com/business/environment/un-chief-convene-no-nonsense-climate-summit-2023-2022-12-19/.
38. Alastair Marsh, "Bankers Told They Can Ignore Binding Fossil-Finance Restrictions," Bloomberg, October 17, 2022, https://www.bloomberg.com/news/articles/2022-10-17/bankers-told-they-can-ignore-binding-fossil-finance-restrictions; Paul C. Stern, Thomas Dietz, and Michael P. Vandenbergh, "The Science of Mitigation: Closing the Gap Between Potential and Actual Reduction of Environmental Threats," *Energy Research & Social Science* 91 (September 1, 2022): 102735, https://doi.org/10.1016/j.erss.2022.102735; David G. Victor, Marcel Lumkowsky, and Astrid Dannenberg, "Determining the Credibility of Commitments in International Climate Policy," *Nature Climate Change* 12, no. 9 (September 2022): 793–800, https://doi.org/10.1038/s41558-022-01454-x.
39. "World on 'Fast Track to Climate Disaster,' Says UN Secretary General: Video," April 4, 2022, https://www.theguardian.com/environment/video/2022/apr/04/world-on-fast-track-to-climate-disaster-say-un-secretary-general-video.
40. See especially Erick Lachapelle and Matthew Paterson, "Drivers of National Climate Policy," *Climate Policy* 13, no. 5 (September 1, 2013): 547–71, https://doi.org/10.1080/14693062.2013.811333; Matthew Lockwood et al., "Historical Institutionalism and the Politics of Sustainable Energy Transitions: A Research Agenda," *Environment and Planning C: Politics and Space* 35, no. 2 (March 1, 2017): 312–33, https://doi.org/10.1177/0263774X16660561; Hugh Ward, Xun Cao, and Bumba Mukherjee,

"State Capacity and the Environmental Investment Gap in Authoritarian States," *Comparative Political Studies* 47, no. 3 (March 1, 2014): 309–43, https://doi.org/10.1177/0010414013509569.

41. Intergovernmental Panel on Climate Change Working Group 3, *Climate Change 2022: Mitigation of Climate Change.*.
42. Climate Action Tracker, USA, accessed June 4, 2023, https://climateactiontracker.org/countries/usa/.
43. Joanna Derman, "The Inflation Reduction Act: Topline Oil and Gas Reforms," Project on Government Oversight, September 2, 2022, https://www.pogo.org/analysis/2022/09/the-inflation-reduction-act-topline-oil-and-gas-reforms.
44. Climate Action Tracker, USA, accessed October 10, 2022, https://climateactiontracker.org/countries/usa/.
45. See discussion by Victor, Lumkowsky, and Dannenberg, "Determining the Credibility of Commitments."
46. Shelby Webb, Carlos Iaconangelo, and David Anchondo, "3 Questions Answered on the Ukraine War's Impact on Energy," E&E News, February 22, 2023, https://www.eenews.net/articles/3-questions-answered-on-the-ukraine-wars-impact-on-energy/.
47. Richard Elliott Benedick, *Ozone Diplomacy* (Cambridge, MA: Harvard University Press, 1998), https://www.hup.harvard.edu/catalog.php?isbn=9780674650039. For an alternative perspective, see Karen Litfin, *Ozone Discourses: Science and Politics in Global Environmental Cooperation* (New York: Columbia University Press, 1994). See also Max Oelschlaeger, "The Myth of the Technological Fix," *The Southwestern Journal of Philosophy* 10, no. 1 (1979): 43–53, https://www.jstor.org/stable/43155445 for a critical assessment.
48. G. Supran, S. Rahmstorf, and N. Oreskes, "Assessing ExxonMobil's Global Warming Projections," *Science* 379, no. 6628 (January 13, 2023): eabk0063, https://doi.org/10.1126/science.abk0063.
49. See, for example, Wil Burns, David Dana, and Simon James Nicholson, eds., *Climate Geoengineering: Science, Law and Governance*, AESS Interdisciplinary Environmental Studies and Sciences Series (Cham, Switzerland: Springer International, 2021), https://doi.org/10.1007/978-3-030-72372-9.
50. Corbin Hiar and Carlos Anchondo, "DOE Releases Record Funding for Removing Carbon," E&E News, December 14, 2022, https://www.eenews.net/articles/doe-releases-record-funding-for-removing-carbon/;

Justin Jacobs, "Oil Companies Line Up for Billions of Dollars in Subsidies Under US Climate Law," *Financial Times*, March 7, 2023, sec. Oil & Gas Industry, https://www.ft.com/content/28b3a8d9-9c5f-4578-a6c6-7b848b3fe700. For a broader overview, see Simon Nicholson and Jesse L. Reynolds, "Taking Technology Seriously: Introduction to the Special Issue on New Technologies and Global Environmental Politics," *Global Environmental Politics* 20, no. 3 (August 1, 2020): 1–8, https://doi.org/10.1162/glep_e_00576.

51. For more detail, see Edgar G. Hertwich and Richard Wood, "The Growing Importance of Scope 3 Greenhouse Gas Emissions from Industry," *Environmental Research Letters* 13, no. 10 (October 2018): 104013, https://doi.org/10.1088/1748-9326/aae19a.
52. Lottie Limb, "'Pure Climate Vandalism': Shell Backtracks on Plans to Cut Oil," Euronews, June 15, 2023, https://www.euronews.com/green/2023/06/15/shell-joins-bp-and-total-in-u-turning-on-climate-pledges-to-reward-shareholders; Tom Wilson and Emma Dunkley, "BP Slows Oil and Gas Retreat After Record $28bn Profit," *Financial Times*, February 7, 2023, https://www.ft.com/content/419f137c-3a83-4c9c-9957-34b6609bcdf7.
53. Jeff Tollefson, "Top Climate Scientists Are Sceptical That Nations Will Rein in Global Warming," *Nature* 599, no. 7883 (November 1, 2021): 22–24, https://doi.org/10.1038/d41586-021-02990-w.
54. Adam Vaughan, "The World's 1.5°C Climate Goal Is Slipping Out of Reach: So Now What?," New Scientist, June 7, 2022, https://www.newscientist.com/article/2323175-the-worlds-1-5c-climate-goal-is-slipping-out-of-reach-so-now-what/. See also Anita Engels et al., "Hamburg Climate Futures Outlook: The Plausibility of a 1.5°C Limit to Global Warming—Social Drivers and Physical Processes," Universität Hamburg, February 1, 2023, https://doi.org/10.25592/uhhfdm.11230.
55. Jonathan W. Kuyper, Björn-Ola Linnér, and Heike Schroeder, "Non-State Actors in Hybrid Global Climate Governance: Justice, Legitimacy, and Effectiveness in a Post-Paris Era," *WIREs Climate Change* 9, no. 1 (2018): e497, https://doi.org/10.1002/wcc.497; Joost de Moor et al., "New Kids on the Block: Taking Stock of the Recent Cycle of Climate Activism," *Social Movement Studies*, October 28, 2020, 1–7, https://doi.org/10.1080/14742837.2020.1836617.
56. Olivia Rosane, "7.6 Million Join Week of Global Climate Strikes," Eco Watch, September 30, 2019, https://www.ecowatch.com/global-climate-strikes-week-2640790405.html. See also Dana R. Fisher and Sohana

Nasrin, "Climate Activism and Its Effects," Wiley Interdisciplinary Reviews: Climate Change, October 18, 2020, e683, https://doi.org/10.1002/wcc.683.

57. For an overview of the radical flank effect, see Herbert H. Haines, "Black Radicalization and the Funding of Civil Rights: 1957–1970," *Social Problems* 32, no. 1 (October 1, 1984): 31–43, https://doi.org/10.2307/800260. See also Brent Simpson, Robb Willer, and Matthew Feinberg, "Radical Flanks of Social Movements Can Increase Support for Moderate Factions," *PNAS Nexus* 1, no. 3 (July 1, 2022): 110, https://doi.org/10.1093/pnasnexus/pgac110; Dana R. Fisher, "Understanding the Growing Radical Flank of the Climate Movement as the World Burns," Brookings Institution, September 2, 2023, https://www.brookings.edu/articles/understanding-the-growing-radical-flank-of-the-climate-movement-as-the-world-burns/.
58. Shannon Osaka, "From Crashing 'The View' to Tomato Soup: Climate Protests Get Weird," *Washington Post*, October 24, 2022, https://www.washingtonpost.com/climate-environment/2022/10/14/tomato-soup-sunflowers-climate-protest/.
59. For a full discussion, see Fisher and Nasrin, "Climate Activism and Its Effects."
60. Fisher and Nasrin, 6.
61. Gert Spaargaren and Arthur P. J. Mol, "Sociology, Environment, and Modernity: Ecological Modernization as a Theory of Social Change," *Society & Natural Resources* 5, no. 4 (October 1, 1992): 323–44, https://doi.org/10.1080/08941929209380797. See also Dana R. Fisher and William R. Freudenburg, "Ecological Modernization and Its Critics: Assessing the Past and Looking Toward the Future," *Society & Natural Resources* 14, no. 8 (2001): 701–9.
62. Anthony Giddens, *The Third Way* Cambridge: Polity, 1991). See also Dana R. Fisher and William R. Freudenburg, "Postindustrialization and Environmental Quality: An Empirical Analysis of the Environmental State," *Social Forces* 83, no. 1 (2004): 157–88; Andrew K. Jorgenson and Brett Clark, "Are the Economy and the Environment Decoupling? A Comparative International Study, 1960–2005," *American Journal of Sociology* 118, no. 1 (2012): 1–44, https://doi.org/10.1086/665990; Anthony Giddens, *The Politics of Climate Change* (Cambridge: Polity, 2009).
63. Naomi Oreskes and Erik M. Conway, *Merchants of Doubt: How a Handful of Scientists Obscured the Truth on Issues from Tobacco Smoke to Global Warming* (New York: Bloomsbury, 2011); Justin Farrell, "Corporate

Funding and Ideological Polarization About Climate Change," *Proceedings of the National Academy of Sciences* 113, no. 1 (January 5, 2016): 92–97, https://doi.org/10.1073/pnas.1509433112.

64. See, for example, Nafeez Ahmed, "IPCC Reports 'Diluted' Under 'Political Pressure' to Protect Fossil Fuel Interests," *Guardian*, May 15, 2014, sec. Environment, https://www.theguardian.com/environment/earth-insight/2014/may/15/ipcc-un-climate-reports-diluted-protect-fossil-fuel-interests; Seth Borenstein, "UN Climate Chief Calls Fossil Fuel Phase Out Key to Curbing Warming but May Not Be on Talks' Agenda," AP News, June 5, 2023, https://apnews.com/article/climate-change-fossil-fuels-negotiations-warming-04be11b9ba7f82ed4b8d5718a56970bb.
65. Phoebe Cooke, "IPCC Report Calls Out 'Vested Interests' Delaying Climate Action," DeSmog (blog), February 28, 2022, https://www.desmog.com/2022/02/28/ipcc-report-calls-out-vested-interests-delaying-climate-action/.
66. Energy Mix, "Record Fossil Extraction from Canada, U.S., Norway Despite Fervent Climate Pledges," The Energy Mix (blog), February 2, 2022, https://www.theenergymix.com/2022/02/02/record-fossil-extraction-from-canada-u-s-norway-despite-fervent-climate-pledges/.
67. See discussion in Christine Ayala, "Former Navy Secretary: The Addiction to Fossil Fuels Empowers Putin," The Hill, March 2, 2022, https://thehill.com/opinion/energy-environment/596336-former-navy-secretary-the-addiction-to-fossil-fuels-empowers-putin.
68. Climate Action Tracker, Norway, accessed February 7, 2023, https://climateactiontracker.org/countries/norway/.
69. Nerijus Adomaitis and Gwladys Fouce, "Norway Plans to Offer Record Number of Arctic Oil, Gas Exploration Blocks," Reuters, January 24, 2023, https://www.reuters.com/business/energy/norway-offers-up-92-new-oil-gas-exploration-blocks-2023-01-24/.
70. Craig Callender, "Fossil-Fuel Money Is Warping Climate Research," *Chronicle of Higher Education*, September 29, 2022, https://www.chronicle.com/article/fossil-fuel-money-is-warping-climate-research; Amy Westervelt, "Universities Need to Break Their Addiction to Fossil Funding" Drilled News, October 18, 2022, https://www.drilledpodcast.com/universities-need-to-break-their-addiction-to-fossil-funding/. See also Fossil Free Research, https://fossilfreeresearch.com/resources/.
71. Geoffrey Supran and Naomi Oreskes, "Rhetoric and Frame Analysis of ExxonMobil's Climate Change Communications," *One Earth* 4, no. 5

(May 21, 2021): 696–719, https://doi.org/10.1016/j.oneear.2021.04.014; Naomi Oreskes, "Why Didn't They Act?," in *The Climate Book*, ed. Greta Thunberg (London: Penguin Random House, 2022), 29–31; Supran, Rahmstorf, and Oreskes, "Assessing ExxonMobil's Global Warming Projections."

72. Jorgenson and Clark, "Are the Economy and the Environment Decoupling?;" Richard York, Eugene A. Rosa, and Thomas Dietz, "Footprints on the Earth: The Environmental Consequences of Modernity," *American Sociological Review* 68, no. 2 (2003): 279–300, https://doi.org/10.2307/1519769.
73. United Nations Environment Programme (UNEP), "Emissions Gap Report 2022" (Nairobi, Kenya: UNEP, October 21, 2022), http://www.unep.org/resources/emissions-gap-report-2022.
74. Christopher Wright and Daniel Nyberg, *Climate Change, Capitalism, and Corporations* (Cambridge: Cambridge University Press, 2015). See also Naomi Klein, *This Changes Everything: Capitalism Vs. the Climate* (New York: Simon and Schuster, 2015).
75. Recording of the event available at https://t.co/TLMuxkBOJw (accessed October 2, 2022).
76. As quoted in Amy Westervelt, "The Market Won't Save Us, Unless We Radically Change It," Hot Take, October 2, 2022, https://www.hottakepod.com/the-market-wont-save-us-unless-we-radically-change-it/. See also Oreskes, "Why Didn't They Act?;" Naomi Oreskes and Erik Conway, "The True Cost of the 'Free' Market Was Exposed by the Pandemic and Climate Change," *Time*, February 28, 2023, https://time.com/6258540/true-cost-of-the-free-market/; Naomi Oreskes and Erik M. Conway, *The Big Myth: How American Business Taught Us to Loathe Government and Love the Free Market* (New York: Bloomsbury, 2023).
77. Kimberly A. Nicholas, *Under the Sky We Make: How to Be Human in a Warming World* (New York: Putnam's Sons, 2021).
78. Carbon Brief, "UNEP: Meeting Global Climate Goals Now Requires 'Rapid Transformation of Societies,'" Carbon Brief, October 27, 2022, https://www.carbonbrief.org/unep-meeting-global-climate-goals-now-requires-rapid-transformation-of-societies/.
79. See, for example, Matto Mildenberger, *Carbon Captured: How Business and Labor Control Climate Politics* (Cambridge, MA: MIT Press, 2020).
80. O. Rueda et al., "Negative-Emissions Technology Portfolios to Meet the 1.5°C Target," *Global Environmental Change* 67 (March 1, 2021): 102238, https://doi.org/10.1016/j.gloenvcha.2021.102238.

81. See, for example, Charles Tilly, "War Making and State Making as Organized Crime," in *Bringing the State Back In*, ed. Peter B. Evans, Dietrich Rueschemeyer, and Theda Skocpol (Cambridge: Cambridge University Press, 1985), 169–91, https://doi.org/10.1017/CBO9780511628283; Karl Polanyi, *The Great Transformation: The Political and Economic Origins of Our Time* (Boston: Beacon, 2001); Sandra Halperin, *War and Social Change in Modern Europe: The Great Transformation Revisited* (Cambridge: Cambridge University Press, 2004); Birkland, *Lessons of Disaster*; Agrawal, "A Positive Side of Disaster." But see also Tiffany H. Morrison et al., "Radical Interventions for Climate-Impacted Systems," *Nature Climate Change* 12, no. 12 (December 2022): 1100–1106, https://doi.org/10.1038/s41558-022-01542-y.
82. See particularly Intergovernmental Panel on Climate Change Working Group 2, *Climate Change 2022: Impacts, Adaptation, and Vulnerability*, chap. 12. See also Luke Kemp et al., "Climate Endgame: Exploring Catastrophic Climate Change Scenarios," *Proceedings of the National Academy of Sciences* 119, no. 34 (August 23, 2022): e2108146119, https://doi.org/10.1073/pnas.2108146119; Daniel Steel, C. Tyler DesRoches, and Kian Mintz-Woo, "Climate Change and the Threat to Civilization," *Proceedings of the National Academy of Sciences* 119, no. 42 (October 18, 2022): e2210525119, https://doi.org/10.1073/pnas.2210525119.
83. Erica Chenoweth and Maria J. Stephan, *Why Civil Resistance Works: The Strategic Logic of Nonviolent Conflict* (New York: Columbia University Press, 2011). See also Erica Chenoweth, "Questions, Answers, and Some Cautionary Updates Regarding the 3.5% Rule," Carr Center Discussion Paper, Carr Center for Human Rights Policy, Harvard Kennedy School, April 2020, https://carrcenter.hks.harvard.edu/files/cchr/files/CCDP_005.pdf.
84. See particularly Andreas Malm, *How to Blow Up a Pipeline: Learning to Fight in a World on Fire* (London: Verso, 2021). See also review of tactics by May Aye Thiri et al., "How Social Movements Contribute to Staying Within the Global Carbon Budget: Evidence from a Qualitative Meta-Analysis of Case Studies," *Ecological Economics* 195 (May 1, 2022): 107356, https://doi.org/10.1016/j.ecolecon.2022.107356.
85. Joshua Bloom, "The Dynamics of Repression and Insurgent Practice in the Black Liberation Struggle," *American Journal of Sociology* 126, no. 2 (September 1, 2020): 195–259, https://doi.org/10.1086/711672. See also Doug McAdam, *Political Process and the Development of Black Insurgency*,

1930–1970, 2nd ed. (Chicago: University of Chicago Press, 1982); Thiri et al., "How Social Movements Contribute."

86. See, for example, Dana R Fisher, *National Governance and the Global Climate Change Regime* (Lanham, MD: Rowman & Littlefield, 2004); Dana R. Fisher, "Bringing the Material Back In: Understanding the U.S. Position on Climate Change," *Sociological Forum* 21, no. 3 (2006): 467–94, https://www.jstor.org/stable/4540952; Dana R. Fisher, Joseph Waggle, and Philip Leifeld, "Where Does Political Polarization Come From? Locating Polarization Within the US Climate Change Debate," *American Behavioral Scientist* 57, no. 1 (2013): 70–92; Lorien Jasny, Joseph Waggle, and Dana R. Fisher, "An Empirical Examination of Echo Chambers in US Climate Policy Networks," *Nature Climate Change* 5, no. 8 (August 2015): 782–86, https://doi.org/10.1038/nclimate2666; Dana R. Fisher and Philip Leifeld, "The Polycentricity of Climate Policy Blockage," *Climatic Change* 155 (July 5, 2019): 469–87, https://doi.org/10.1007/s10584-019-02481-y; Lorien Jasny and Dana R Fisher, "Echo Chambers in Climate Science," *Environmental Research Communications* 1, no. 10 (October 11, 2019): 101003, https://doi.org/10.1088/2515-7620/ab491c.

2. SAVING OURSELVES IS A LONG GAME: WHY OUR INSTITUTIONS KEEP FAILING TO ACT ON CLIMATE

1. "The World Is Going to Miss the Totemic 1.5°C Climate Target," *Economist*, November 5, 2022, https://www.economist.com/interactive/briefing/2022/11/05/the-world-is-going-to-miss-the-totemic-1-5c-climate-target.
2. Intergovernmental Panel on Climate Change (IPCC) Working Group 1, *Climate Change 2021: The Physical Science Basis. Contribution of Working Group I to the Sixth Assessment Report of the Intergovernmental Panel on Climate Change* (Cambridge: Cambridge University Press, 2021), https://www.ipcc.ch/report/sixth-assessment-report-working-group-i/; Intergovernmental Panel on Climate Change (IPCC) Working Group 2, *Climate Change 2022: Impacts, Adaptation, and Vulnerability. Contribution of Working Group II to the Sixth Assessment Report of the Intergovernmental Panel on Climate Change* (Cambridge: Cambridge University Press, 2022), https://www.ipcc.ch/report/ar6/wg2/; Intergovernmental Panel on Climate Change (IPCC) Working Group 3, *Climate Change 2022: Mitigation*

of Climate Change. Contribution of Working Group III to the Sixth Assessment Report of the Intergovernmental Panel on Climate Change (Cambridge: Cambridge University Press, 2022), doi:10.1017/9781009157926.

3. For a full report, see United Nations Environment Programme (UNEP), "Emissions Gap Report 2022" (Nairobi, Kenya: UNEP, October 21, 2022), http://www.unep.org/resources/emissions-gap-report-2022, accessed March 17, 2022.
4. UN News, "'Cooperate or Perish': At COP27 UN Chief Calls for Climate Solidarity Pact, Urges Tax on Oil Companies to Finance Loss and Damage," UN News, November 7, 2022, https://news.un.org/en/story/2022/11/1130247. See also Fiona Harvey and Damian Carrington, "World Is on 'Highway to Climate Hell,' UN Chief Warns at COP27 Summit," *Guardian*, November 7, 2022, sec. Environment, https://www.theguardian.com/environment/2022/nov/07/cop27-climate-summit-un-secretary-general-antonio-guterres.
5. Climate Action Tracker, "Massive Gas Expansion Risks Overtaking Positive Climate Policies," November 10, 2022, https://climateactiontracker.org/publications/massive-gas-expansion-risks-overtaking-positive-climate-policies/.
6. For an overview of previous efforts, see Dana R. Fisher, *National Governance and the Global Climate Change Regime* (Lanham, MD: Rowman & Littlefield, 2004); Dana R. Fisher, "Bringing the Material Back In: Understanding the U.S. Position on Climate Change," *Sociological Forum* 21, no. 3 (2006): 467–94, https://www.jstor.org/stable/4540952; Barry Rabe, "Environmental Policy and the Bush Era: The Collision Between the Administrative Presidency and State Experimentation," *Publius: The Journal of Federalism* 37, no. 3 (January 1, 2007): 413–31, https://doi.org/10.1093/publius/pjm007; Lorien Jasny, Joseph Waggle, and Dana R. Fisher, "An Empirical Examination of Echo Chambers in US Climate Policy Networks," *Nature Climate Change* 5, no. 8 (August 2015): 782–86, https://doi.org/10.1038/nclimate2666; Dana R. Fisher and Philip Leifeld, "The Polycentricity of Climate Policy Blockage," *Climatic Change* 155 (2019): 469–87, https://doi.org/10.1007/s10584-019-02481-y.
7. Kate Aronoff, "The Manchin Climate Deal Is Both a Big Win and a Deal with the Devil," *The New Republic*, August 3, 2022, https://newrepublic.com/article/167272/manchin-climate-deal-big-win-deal-devil. See also Joanna Derman, "The Inflation Reduction Act: Topline Oil and Gas

Reforms," Project On Government Oversight, September 2, 2022, https://www.pogo.org/analysis/2022/09/the-inflation-reduction-act-topline-oil-and-gas-reforms.

8. David G. Victor, Marcel Lumkowsky, and Astrid Dannenberg, "Determining the Credibility of Commitments in International Climate Policy," *Nature Climate Change* 12, no. 9 (September 2022): 793–800, https://doi.org/10.1038/s41558-022-01454-x; Paul C. Stern, Thomas Dietz, and Michael P. Vandenbergh, "The Science of Mitigation: Closing the Gap Between Potential and Actual Reduction of Environmental Threats," *Energy Research & Social Science* 91 (September 1, 2022): 102735, https://doi.org/10.1016/j.erss.2022.102735.
9. United Nations Environment Programme (UNEP), *Emissions Gap Report 2022*, October 27, 2022, https://www.unep.org/resources/emissions-gap-report-2022.
10. For a discussion of how the regime has been insufficient in addressing the problem, see Isak Stoddard et al., "Three Decades of Climate Mitigation: Why Haven't We Bent the Global Emissions Curve?," *Annual Review of Environment and Resources* 46, no. 1 (2021): 653–89, https://doi.org/10.1146/annurev-environ-012220-011104.
11. Simon Evans, "Which Countries Are Historically Responsible for Climate Change?," Carbon Brief, October 5, 2021, https://www.carbonbrief.org/analysis-which-countries-are-historically-responsible-for-climate-change/.
12. See, for example, Timothy Puko and Steven Mufson, "Arctic Oil Project Was a Conundrum for Biden: Will There Be Others?," *Washington Post*, March 14, 2023, https://www.washingtonpost.com/climate-environment/2023/03/13/willow-project-alaska-biden-conocophillips/.
13. This list is not meant to be exhaustive. For an overview, see United Nations Conference on Environment and Development, June 3–14, 1992, "A New Blueprint for International Action on the Environment," accessed November 8,2022, https://www.un.org/en/conferences/environment/rio1992.
14. For an overview, see United Nations Climate Change, "What Is the United Nations Framework Convention on Climate Change?," accessed December 4, 2022, https://unfccc.int/process-and-meetings/what-is-the-united-nations-framework-convention-on-climate-change.
15. United Nations Climate Change, "What Is the Kyoto Protocol?," accessed November 5, 2022, https://unfccc.int/kyoto_protocol.

16. For a full discussion, see Fisher, *National Governance and the Global Climate Change Regime*.
17. International Institute for Sustainable Development (IISD), "UNFCCC Reports on Annex I National GHG Inventory Data for 1990–2012," November 24, 2014, http://sdg.iisd.org/news/unfccc-reports-on-annex-i-national-ghg-inventory-data-for-1990-2012/.
18. "Inventory of U.S. Greenhouse Gas Emissions and Sinks: 1990–2012," Environmental Protection Agency, April 15, 2014, https://www.epa.gov/ghgemissions/inventory-us-greenhouse-gas-emissions-and-sinks-1990-2012.
19. Ed King, "Kyoto Protocol: 10 Years of the World's First Climate Change Treaty," Climate Home News, February 16, 2015, https://www.climatechangenews.com/2015/02/16/kyoto-protocol-10-years-of-the-worlds-first-climate-change-treaty/. See also "Canada Pulls Out of Kyoto Protocol," *Guardian*, December 13, 2011, sec. Environment, https://www.theguardian.com/environment/2011/dec/13/canada-pulls-out-kyoto-protocol.
20. "2012's Carbon Emissions in Five Graphs," Carbon Brief, November 1, 2013, https://www.carbonbrief.org/2012s-carbon-emissions-in-five-graphs/.
21. Victor, Lumkowsky, and Dannenberg, "Determining the Credibility of Commitments."
22. United States Environmental Protection Agency, "Fact Sheet: Overview of the Clean Power Plan: Cutting Carbon Pollution from Power Plants, last updated on May 9, 2017, https://archive.epa.gov/epa/cleanpowerplan/fact-sheet-overview-clean-power-plan.html. See, for example, Lorien Jasny et al., "Shifting Echo Chambers in US Climate Policy Networks," *PLOS ONE* 13, no. 9 (September 14, 2018): e0203463, https://doi.org/10.1371/journal.pone.0203463. See also Fisher and Leifeld, "The Polycentricity of Climate Policy Blockage."
23. For details, see Fisher, *National Governance and the Global Climate Change Regime*.
24. United Nations Treaty Collections, Chapter XXVII: Environment, 7, d Paris Agreement, December 12, 2015, status as at March 8, 3023, https://treaties.un.org/Pages/ViewDetails.aspx?src=TREATY&mtdsg_no=XXVII-7-d&chapter=27&clang=_en#5.
25. United Nations Climate Change, "The Paris Agreement: What Is the Paris Agreement?," accessed November 5, 2022, https://unfccc.int/process-and-meetings/the-paris-agreement/the-paris-agreement.

26. Intergovernmental Panel on Climate Change (IPCC) Working Group 1, *Climate Change 2021: The Physical Science Basis.*
27. International Energy Agency (IEA), "Global CO2 Emissions Rebounded to Their Highest Level in History in 2021," press release, March 8, 2022, https://www.iea.org/news/global-co2-emissions-rebounded-to-their-highest-level-in-history-in-2021.
28. Fergus Green and Harro van Asselt, "COP27 Flinched on Phasing Out 'All Fossil Fuels': What's Next for the Fight to Keep Them in the Ground?," The Conversation, November 22, 2022, http://theconversation.com/cop27-flinched-on-phasing-out-all-fossil-fuels-whats-next-for-the-fight-to-keep-them-in-the-ground-194941; Joe Lo and Chloe Farand, "Late-Night Fossil Fuel Fight Leaves Bitter Taste after Cop27," Climate Home News, November 24, 2022, https://www.climatechangenews.com/2022/11/24/late-night-fossil-fuel-fight-leaves-bitter-taste-after-cop27/.
29. "Climate Change: Five Key Takeaways from COP27," *BBC News*, November 20, 2022, sec. Science & Environment, https://www.bbc.com/news/science-environment-63693738.
30. Lo and Farand, "Late-Night Fossil Fuel Fight."
31. Jocelyn Timperley, "The Broken $100-Billion Promise of Climate Finance—and How to Fix It," *Nature* 598, no. 7881 (October 20, 2021): 400–2, https://doi.org/10.1038/d41586-021-02846-3. See also Oxfam International, "Rich Countries' Continued Failure to Honor Their $100 Billon Climate Finance Promise Threatens Negotiations and Undermines Climate Action," press release, June 5, 2023, https://www.oxfam.org/en/press-releases/rich-countries-continued-failure-honor-their-100-billon-climate-finance-promise.
32. For an overview, see Harriet Bulkeley et al., *Transnational Climate Change Governance* (New York: Cambridge University Press, 2014); Daniel H. Cole, "Advantages of a Polycentric Approach to Climate Change Policy," *Nature Climate Change* 5, no. 2 (February 2015): 114–18, https://doi.org/10.1038/nclimate2490; Andrew J. Jordan et al., "Emergence of Polycentric Climate Governance and Its Future Prospects," *Nature Climate Change* 5, no. 11 (November 2015): 977–82, https://doi.org/10.1038/nclimate2725. But see Fisher and Leifeld, "The Polycentricity of Climate Policy Blockage."
33. "World of Change: Global Temperatures," Earth Observatory, accessed October 11, 2022, https://earthobservatory.nasa.gov/world-of-change/global-temperatures.

34. See, for example, Chen Zhou et al., "Greater Committed Warming After Accounting for the Pattern Effect," *Nature Climate Change* 11, no. 2 (February 2021): 132–36, https://doi.org/10.1038/s41558-020-00955-x.
35. Jeff Tollefson, "Top Climate Scientists Are Sceptical That Nations Will Rein in Global Warming," *Nature* 599, no. 7883 (November 1, 2021): 22–24, https://doi.org/10.1038/d41586-021-02990-w.
36. See, for example, Anita Engels et al., "Hamburg Climate Futures Outlook: The Plausibility of a 1.5°C Limit to Global Warming—Social Drivers and Physical Processes," Universität Hamburg, February 1, 2023, https://doi.org/10.25592/uhhfdm.11230.
37. Dana R. Fisher, "COP-15 in Copenhagen: How the Merging of Movements Left Civil Society Out in the Cold," *Global Environmental Politics* 10, no. 2 (2010): 11–17. For an historical perspective, see Dana R. Fisher and Jessica F. Green, "Understanding Disenfranchisement: Civil Society and Developing Countries' Influence and Participation in Global Governance for Sustainable Development," *Global Environmental Politics* 4, no. 3 (2004): 65–84.
38. See, for example, Naomi Klein, "Holding the COP27 Summit in Egypt's Police State Creates a Moral Crisis for the Climate Movement," The Intercept, October 7, 2022, https://theintercept.com/2022/10/07/egypt-cop27-climate-prisoners-alaa/.
39. "Over 25 Percent More Fossil Fuel Lobbyists Than Last Year, Flooding Crucial COP Climate Talks," Corporate Accountability, November 10, 2022, https://www.corporateaccountability.org/media/fossil-fuel-lobbyists-cop27/.
40. Damian Carrington, "The 1.5C Climate Goal Died at Cop27—but Hope Must Not," *Guardian*, November 20, 2022, sec. Environment, https://www.theguardian.com/environment/2022/nov/20/cop27-summit-climate-crisis-global-heating-fossil-fuel-industry.
41. See "Warning Stripes," Institute for Environmental Analytics, accessed November 23, 2022, https://showyourstripes.info/s/globe.
42. Naomi Oreskes, "Why Didn't They Act?," in *The Climate Book*, ed. Greta Thunberg (London: Penguin Random House, 2022), 29–30. See also G. Supran, S. Rahmstorf, and N. Oreskes, "Assessing ExxonMobil's Global Warming Projections," *Science* 379, no. 6628 (January 13, 2023): eabk0063, https://doi.org/10.1126/science.abk0063.
43. For an overview, see Fisher, "Bringing the Material Back In;" Rabe, "Environmental Policy and the Bush Era." See also Fisher and Leifeld, "The Polycentricity of Climate Policy Blockage."

44. Gert Spaargaren and Arthur P. J. Mol, "Sociology, Environment, and Modernity: Ecological Modernization as a Theory of Social Change," *Society & Natural Resources* 5, no. 4 (October 1, 1992): 323–44, https://doi.org/10.1080/08941929209380797; Arthur P. J. Mol and M. Janicke, "The Origins and Theoretical Foundations of Ecological Modernization Theory," in *The Ecological Modernisation Reader: Environmental Reform in Theory and Practice*, ed. Arthur P. J. Mol, David A Sonnenfeld, and Gert Spaargaren (London: Routledge, 2009), 17–27; Bill Gates, *How to Avoid a Climate Disaster: The Solutions We Have and the Breakthroughs We Need* (New York: Knopf, 2021).
45. William R. Freudenburg, "Privileged Access, Privileged Accounts: Toward a Socially Structured Theory of Resources and Discourses," *Social Forces* 84, no. 1 (2005): 89–114. See also Don Grant, Andrew Jorgenson, and Wesley Longhofer, *Super Polluters: Tackling the World's Largest Sites of Climate-Disrupting Emissions* (New York: Columbia University Press, 2020).
46. William R. Freudenburg and Robert Gramling, "Bureaucratic Slippage and Failures of Agency Vigilance: The Case of the Environmental Studies Program," *Social Problems* 41, no. 2 (1994): 214–39; Lindsey Dillon et al., "The Environmental Protection Agency in the Early Trump Administration: Prelude to Regulatory Capture," *American Journal of Public Health* 108, no. S2 (April 1, 2018): S89–S94, https://doi.org/10.2105/AJPH.2018.304360; Alexander Hertel-Fernandez, *State Capture: How Conservative Activists, Big Businesses, and Wealthy Donors Reshaped the American States—and the Nation* (Oxford: Oxford University Press, 2019).
47. Peter J. Jacques, Riley E. Dunlap, and Mark Freeman, "The Organisation of Denial: Conservative Think Tanks and Environmental Scepticism," *Environmental Politics* 17, no. 3 (June 1, 2008): 349–85, https://doi.org/10.1080/09644010802055576; Justin Farrell, "Corporate Funding and Ideological Polarization About Climate Change," *Proceedings of the National Academy of Sciences* 113, no. 1 (January 5, 2016): 92–97, https://doi.org/10.1073/pnas.1509433112; Justin Farrell, "The Growth of Climate Change Misinformation in US Philanthropy: Evidence from Natural Language Processing," *Environmental Research Letters* 14, no. 3 (March 2019): 034013, https://doi.org/10.1088/1748-9326/aaf939; Geoffrey Supran and Naomi Oreskes, "Assessing ExxonMobil's Climate Change Communications (1977–2014)," *Environmental Research Letters* 12, no. 8 (2017): 084019, https://doi.org/10.1088/1748-9326/aa815f.

48. Adele Peters, "These Groups Fighting Offshore Wind Say It's About Whales—but They're Funded by Big Oil," Fast Company, March 1, 2023, https://www.fastcompany.com/90856401/these-groups-fighting-offshore-wind-say-its-about-whales-but-theyre-funded-by-big-oil.
49. See, for example, Nafeez Ahmed, "IPCC Reports 'Diluted' Under 'Political Pressure' to Protect Fossil Fuel Interests," *Guardian*, May 15, 2014, sec. Environment, https://www.theguardian.com/environment/earth-insight/2014/may/15/ipcc-un-climate-reports-diluted-protect-fossil-fuel-interests; Phoebe Cooke, "IPCC Report Calls Out 'Vested Interests' Delaying Climate Action," DeSmog (blog), February 28, 2022, https://www.desmog.com/2022/02/28/ipcc-report-calls-out-vested-interests-delaying-climate-action/.
50. Corbin Hiar and Carlos Anchondo, "DOE Releases Record Funding for Removing Carbon," E&E News, December 14, 2022, https://www.eenews.net/articles/doe-releases-record-funding-for-removing-carbon/; Justin Jacobs, "Oil Companies Line up for Billions of Dollars in Subsidies Under US Climate Law," *Financial Times*, March 7, 2023, sec. Oil & Gas industry, https://www.ft.com/content/28b3a8d9-9c5f-4578-a6c6-7b848b3fe700.
51. Justin Guay, "Closing the Clean Investment Gap," Energy Monitor (blog), November 9, 2022, https://www.energymonitor.ai/finance/opinion-closing-the-clean-investment-gap.
52. Available at "2023 Statistical Review of World Energy," Energy Institute, accessed June 26, 2023, https://www.energyinst.org/statistical-review.
53. U.S. Energy Information Administration (EIA) "Monthly Crude Oil and Natural Gas Production," July 31, 2023, https://www.eia.gov/petroleum/production/#ng-tab.
54. Puko and Mufson, "Arctic Oil Project Was a Conundrum for Biden." See also Kate Aronoff, "Why Is the Fossil Fuel Industry Praising the Inflation Reduction Act?," *The New Republic*, March 10, 2023, https://newrepublic.com/article/171086/fossil-fuel-industry-praising-inflation-reduction-act.
55. Edgar G. Hertwich and Richard Wood, "The Growing Importance of Scope 3 Greenhouse Gas Emissions from Industry," *Environmental Research Letters* 13, no. 10 (October 2018): 104013, https://doi.org/10.1088/1748-9326/aae19a.
56. "COP27: 'Zero Tolerance for Greenwashing', Guterres Says as New Report Cracks Down on Empty Net-Zero Pledges," UN News, November 8, 2022, https://news.un.org/en/story/2022/11/1130317.

57. Tom Wilson and Emma Dunkley, "BP Slows Oil and Gas Retreat After Record $28bn Profit," *Financial Times*, February 7, 2023, https://www.ft.com/content/419f137c-3a83-4c9c-9957-34b6609bcdf7; Stuart Braun, "Shell, BP Boost Profit, Sink Investment in Renewable Energy," DW, February 10, 2023, https://www.dw.com/en/shell-bp-boost-profit-sink-investment-in-renewable-energy/a-64656800; Lottie Limb, "'Pure Climate Vandalism': Shell Backtracks on Plans to Cut Oil," Euronews, June 15, 2023, https://www.euronews.com/green/2023/06/15/shell-joins-bp-and-total-in-u-turning-on-climate-pledges-to-reward-shareholders.
58. Fisher, "Bringing the Material Back In."
59. Evans, "Which Countries Are Historically Responsible for Climate Change?"
60. In his opening comments, Joseph Aldy listed the following climate 'roadkill': McCain-Lieberman, Waxman-Markey, the Clean Power Plan, the Green New Deal, and the Baker-Shultz Carbon Dividends Plan.
61. For a copy of the letter announcing his formal position, see White House, Office of the Press Secretary, "Text of a Letter from the President to Senators, Hagel, Helms, Craig, and Roberts," March 13, 2001, https://georgewbush-whitehouse.archives.gov/news/releases/2001/03/20010314.html.
62. See, for example, Fisher, "Bringing the Material Back In."
63. For a full discussion, see Fisher, *National Governance and the Global Climate Change Regime*, chap. 6.
64. Matthew H. Goldberg et al., "Oil and Gas Companies Invest in Legislators That Vote Against the Environment," *Proceedings of the National Academy of Sciences* 117, no. 10 (2020): 5111–12, https://doi.org/10.1073/pnas.1922175117.
65. Fisher, "Bringing the Material Back In."
66. Jasny, Waggle, and Fisher, "An Empirical Examination of Echo Chambers."
67. Jasny et al., "Shifting Echo Chambers in US Climate Policy Networks;" Lorien Jasny and Dana R. Fisher, "Echo Chambers in Climate Science," *Environmental Research Communications* 1, no. 10 (October 11, 2019): 101003, https://doi.org/10.1088/2515-7620/ab491c; Fisher and Leifeld, "The Polycentricity of Climate Policy Blockage."
68. Steve Peoples, "In Intimate Moment, Biden Vows to 'End Fossil Fuel,'" Associated Press, September 6, 2019, https://apnews.com/article/9dfb1e4c381043bab6fd0fa6dece3974.

69. Dana R. Fisher, "Youth Climate Activists Once Opposed Joe Biden: Now, They Say They'll Vote for Him.," Politico, May 8, 2020, https://www.politico.com/news/magazine/2020/05/08/youth-climate-activists-joe-biden-survey-241068.
70. Brady Dennis and Dino Grandoni, "How Joe Biden's Surprisingly Ambitious Climate Plan Came Together," *Washington Post*, August 2, 2020, https://www.washingtonpost.com/climate-environment/how-joe-bidens-surprisingly-ambitious-climate-plan-came-together/2020/07/31/b73e78d0-cd11-11ea-91f1-28aca4d833a0_story.html.
71. Ben Kesslen, "Sunrise Movement Made 1.3 Million Calls, 2.4 Million Texts for Biden, Other Candidates," NBC News, January 6, 2021, https://www.nbcnews.com/politics/2020-election/blog/election-day-2020-live-updates-n1245892.
72. "Budget Reconciliation: The Basics," House Budget Committee Democrats, August 11, 2021, https://budget.house.gov/publications/fact-sheet/budget-reconciliation-basics.
73. Derman, "The Inflation Reduction Act."
74. Aronoff, "The Manchin Climate Deal."
75. Coral Davenport and Brad Plumer, "Debt Deal Includes a Green Light for a Contentious Pipeline," *New York Times*, May 30, 2023, sec. Climate, https://www.nytimes.com/2023/05/30/climate/mountain-valley-pipe.html. See also Aronoff, "The Manchin Climate Deal."
76. Although previous administrations used executive order to craft a policy because they were unable to pass legislation, this is the first climate policy that passed through the Congress and was signed by the president.
77. Fisher, interview with Jason Walsh, February 4, 2022.
78. Climate Action Tracker, USA, accessed February 4, 2023, https://climateactiontracker.org/countries/usa/.
79. Climate Action Tracker estimates that the maximum emissions reductions at around 42 percent below 2005 levels. See also John Larsen et al., "A Turning Point for US Climate Progress: Assessing the Climate and Clean Energy Provisions in the Inflation Reduction Act," Rhodium Group (blog), August 12, 2022, https://rhg.com/research/climate-clean-energy-inflation-reduction-act/; Dan Zukowski, "EV Transition Not Keeping Pace with US Climate Goals: Report," Utility Dive, February 1, 2023, https://www.utilitydive.com/news/ev-electric-vehicles-fleet-electrification-planning-us-climate-goals-Paris-agreement-ICCT/641652/.

80. Leah C. Stokes, "This Year Was the Beginning of a Green Transition," *New York Times*, December 25, 2022, sec. Opinion, https://www.nytimes.com/2022/12/25/opinion/gas-prices-crisis-climate-change.html.
81. See the discussion in Max Zahn, "Biden Climate Law Spurred Billions in Clean Energy Investment: Has It Been a Success?," ABC News, January 29, 2023, https://abcnews.go.com/Business/biden-climate-law-spurred-billions-clean-energy-investment/story?id=96632120. See also Maxine Joselow, "Biden Is Approving More Oil and Gas Drilling Permits on Public Lands Than Trump, Analysis Finds," *Washington Post*, December 8, 2021, https://www.washingtonpost.com/politics/2021/12/06/biden-is-approving-more-oil-gas-drilling-permits-public-lands-than-trump-analysis-finds/.
82. Paul C. Stern et al., "Feasible Climate Mitigation," *Nature Climate Change* 13, no. 1 (December 22, 2022): 6–8, https://doi.org/10.1038/s41558-022-01563-7.
83. For full report, see Dana R. Fisher, "What to Expect on Climate Change from the New Congress," FixGov Brookings Institution (blog), January 3, 2023, https://www.brookings.edu/blog/fixgov/2023/01/03/what-to-expect-on-climate-change-from-the-new-congress/. Because the sample included only two scientists, they were dropped from the figures.
84. Fisher, off-the-record interview with congressional staff, March 2022.
85. Nick Robertson, "House Republicans Blast Environmental Rules in First Energy Meeting," The Hill (blog), February 1, 2023, https://thehill.com/policy/energy-environment/3839521-house-republicans-blast-environmental-rules-in-first-energy-meeting/.
86. Chris Megerian, "Biden Climate Legacy Tested by Backlash over Willow Project," *Washington Post*, March 16, 2023, https://www.washingtonpost.com/politics/2023/03/16/biden-climate-change-willow-alaska-oil-drilling/5c944f8e-c432-11ed-82a7-6a87555c1878_story.html.

3. SAVING OURSELVES INVOLVES TAKING POWER BACK FOR THE PEOPLE

1. Rick Noack, "Europe's Snowless Ski Resorts Preview Winter in a Warming World," *Washington Post*, January 12, 2023, https://www.washingtonpost.com/world/2023/01/12/skiing-climate-change-alps-snow/; Margaret Osbourne, "Skiing Faces an Uncertain Future as Winters Warm,"

Smithsonian Magazine, February 14, 2023, https://www.smithsonianmag.com/smart-news/skiing-faces-an-uncertain-future-as-winters-warm-180981640/.

2. Kasha Patel, "How Climate Change Will Make Atmospheric Rivers Even Worse," *Washington Post*, January 14, 2023, https://www.washingtonpost.com/weather/2023/01/12/climate-change-atmospheric-rivers-rain/. See also National Oceanic & Atmospheric Administration, National Environmental Satellite, Data and Information Service, "Atmospheric Rivers Hit West Coast," January 25, 2023, https://www.nesdis.noaa.gov/news/atmospheric-rivers-hit-west-coast.
3. Chris Mooney, "Floating Ice Around Antarctica Just Hit a Record Low," *Washington Post*, February 15, 2023, https://www.washingtonpost.com/climate-environment/2023/02/14/antarctic-sea-ice-record-low/.
4. Rachel Pannett and Dan Stillman, "New Zealand Declares Rare State of Emergency as Cyclone Wreaks Havoc," *Washington Post*, February 14, 2023, https://www.washingtonpost.com/weather/2023/02/13/new-zealand-cyclone-gabrielle-auckland/.
5. Karen Shih, "2023 UMD Summer Camps: Perennial Favorites and New Programs Offer Choices for Kid of All Ages," Maryland Today, accessed February 23, 2023, https://today.umd.edu/2023-umd-summer-camps.
6. Andrew Freedman, "Pacific Northwest Heat Wave Shatters Early Season Records as Wildfires Burn in Canada," Axios, May 16, 2023, https://www.axios.com/2023/05/16/pacific-northwest-heat-canada-wildfires-climate; Elizabeth Weise and Doyle Rice, "As Canada Burns, Smoke Makes US Air Unhealthy and Skies Eerie. Is Climate Change to Blame?" *USA Today*, June 7, 2023.
7. "Biggest Climate Toll in Year of 'Devastating' Disasters Revealed," *The Guardian*, December 27, 2022, https://www.theguardian.com/environment/2022/dec/27/biggest-climate-toll-in-year-of-devastating-disasters-revealed; Matthew Rozsa, "90 percent of Humans Will Suffer Extreme Heat, Drought Due to Climate Change: Report," Salon, January 10, 2023, https://www.salon.com/2023/01/10/90-of-humans-will-suffer-extreme-heat-drought-due-to-climate-change-report/. For details, see Jiabo Yin et al., "Future Socio-Ecosystem Productivity Threatened by Compound Drought–Heatwave Events," *Nature Sustainability*, January 5, 2023, 1–14, https://doi.org/10.1038/s41893-022-01024-1; Damian Carrington, "World Risks Descending into a Climate 'Doom Loop' Warn Thinktanks," *The Guardian*, February 16, 2023, https://amp-theguardian-com.cdn.ampproject

.org/c/s/amp.theguardian.com/environment/2023/feb/16/world-risks-descending-into-a-climate-doom-loop-warn-thinktanks.

8. For a broad overview, see Jean L. Cohen and Andrew Arato, *Civil Society and Political Theory*, repr. ed. (Cambridge, MA: MIT Press, 1994).
9. See particularly James Bell et al., "In Response to Climate Change, Citizens in Advanced Economies Are Willing to Alter How They Live and Work," Pew Research Center's Global Attitudes Project (blog), September 14, 2021, https://www.pewresearch.org/global/2021/09/14/in-response-to-climate-change-citizens-in-advanced-economies-are-willing-to-alter-how-they-live-and-work/; Anthony Leiserowitz et al., "International Public Opinion on Climate Change, 2022," Yale Program on Climate Change Communication (blog), June 29, 2022, https://climatecommunication.yale.edu/publications/international-public-opinion-on-climate-change-2022/; M. Ballew et al., "Global Warming's Six Americas Across Age, Race/Ethnicity, and Gender," Yale Program on Climate Change Communication (blog), April 5, 2023, https://live-yccc.pantheon.io/publications/global-warmings-six-americas-age-race-ethnicity-gender/.
10. Moira Fagan and Christine Huang, "A Look at How People Around the World View Climate Change," Pew Research Center (blog), April 18, 2019, https://www.pewresearch.org/fact-tank/2019/04/18/a-look-at-how-people-around-the-world-view-climate-change/; Thea Gregersen et al., "Political Orientation Moderates the Relationship Between Climate Change Beliefs and Worry About Climate Change," *Frontiers in Psychology* 11 (July 16, 2020): 1573, https://doi.org/10.3389/fpsyg.2020.01573; Cary Funk, "Key Findings: How Americans' Attitudes About Climate Change Differ by Generation, Party and Other Factors," Pew Research Center (blog), May 26, 2021, https://www.pewresearch.org/fact-tank/2021/05/26/key-findings-how-americans-attitudes-about-climate-change-differ-by-generation-party-and-other-factors/; Taciano L. Milfont et al., "Ten-Year Panel Data Confirm Generation Gap but Climate Beliefs Increase at Similar Rates across Ages," *Nature Communications* 12, no. 1 (July 6, 2021): 4038, https://doi.org/10.1038/s41467-021-24245-y.
11. Reid Wilson, "Climate Change Likely to Hit Red States Hardest," The Hill, January 29, 2021, https://thehill.com/policy/energy-environment/427479-climate-change-likely-to-hit-red-states-hardest/.
12. Brian Eckhouse, "Green Factories Are Changing Minds in More Conservative US States," BNN Bloomberg, November 28, 2022, https://

www.bnnbloomberg.ca/green-factories-are-changing-minds-in-more-conservative-us-states-1.1851815. See also Samuel Trachtman and Jonas Meckling, "The Climate Advocacy Gap," *Climatic Change* 172, no. 3 (June 2, 2022): 24, https://doi.org/10.1007/s10584-022-03381-4.

13. See, for example, Thomas A. Heberlein and J. Stanley Black, "Attitudinal Specificity and the Prediction of Behavior in a Field Setting," *Journal of Personality and Social Psychology* 33 (1976): 474–79, https://doi.org/10.1037/0022-3514.33.4.474.
14. Intergovernmental Panel on Climate Change, *AR6 Synthesis Report*, accessed May 8, 2023, https://www.ipcc.ch/report/ar6/syr/resources/spm-headline-statements.
15. Noah S. Diffenbaugh and Elizabeth A. Barnes, "Data-Driven Predictions of the Time Remaining Until Critical Global Warming Thresholds Are Reached," *Proceedings of the National Academy of Sciences* 120, no. 6 (February 7, 2023): e2207183120, https://doi.org/10.1073/pnas.2207183120.
16. Intergovernmental Panel on Climate Change (IPCC) Working Group 3, *Climate Change 2022: Mitigation of Climate Change. Contribution of Working Group III to the Sixth Assessment Report of the Intergovernmental Panel on Climate Change* (Cambridge: Cambridge University Press, 2022), doi: 10.1017/9781009157926.
17. Smithsonian, National Museum of American History, "Energy Crisis," accessed December 30, 2022, https://americanhistory.si.edu/american-enterprise-exhibition/consumer-era/energy-crisis.
18. Gerald T. Gardner and Paul C. Stern, "The Short List: The Most Effective Actions U.S. Households Can Take to Curb Climate Change," *Environment: Science and Policy for Sustainable Development* 50, no. 5 (September 1, 2008): 12–25, https://doi.org/10.3200/ENVT.50.5.12-25. See also Thomas Dietz et al., "Household Actions Can Provide a Behavioral Wedge to Rapidly Reduce US Carbon Emissions," *Proceedings of the National Academy of Sciences* 106, no. 44 (November 3, 2009): 18452–56, https://doi.org/10.1073/pnas.0908738106.
19. Seth Wynes and Kimberly A. Nicholas, "The Climate Mitigation Gap: Education and Government Recommendations Miss the Most Effective Individual Actions," *Environmental Research Letters* 12, no. 7 (July 2017): 074024, https://doi.org/10.1088/1748-9326/aa7541. See also Seth Wynes et al., "Measuring What Works: Quantifying Greenhouse Gas Emission Reductions of Behavioural Interventions to Reduce Driving, Meat Consumption, and Household Energy Use," *Environmental Research Letters*

13, no. 11 (November 1, 2018): 113002, https://doi.org/10.1088/1748-9326/aae5d7; Kimberly A. Nicholas, *Under the Sky We Make: How to Be Human in a Warming World* (New York: Putnam's Sons, 2021); Kristian S. Nielsen et al., "The Role of High-Socioeconomic-Status People in Locking in or Rapidly Reducing Energy-Driven Greenhouse Gas Emissions," *Nature Energy* 6, no. 11 (November 2021): 1011–16, https://doi.org/10.1038/s41560-021-00900-y.

20. IEA (2022), Norway 2022, IEA, Paris, accessed December 24, 2022, https://www.iea.org/reports/norway-2022, License: CC BY 4.0.
21. Norwegian Government Security and Service Organisation (GSSO), *National Transport Plan 2022–2033*, Meld. St. 20 (2020–2021), Report to the Storting, June 25, 2021, https://www.regjeringen.no/en/dokumenter/national-transport-plan-2022-2033/.
22. Christina Bu, "What Norway Can Teach the World About Switching to Electric Vehicles," *Time*, January 7, 2022, https://time.com/6133180/norway-electric-vehicles/.
23. Martin Thronsen, "Electric Car Sales in 2022: Norway Celebrates Another Record-Breaking Year for Electric Vehicles," Elbil, January 2, 2023, https://elbil.no/norway-celebrates-another-record-breaking-year-for-electric-vehicles/.
24. U.S. Energy Information Administration, "What Is U.S. Electricity Generation by Energy Source?," February 2023, https://www.eia.gov/tools/faqs/faq.php?id=427&t=3.
25. Peter Johnson, "Here's How US Electric Vehicle Sales by Maker and EV Model Through Q3 2022 Compare," Electrek, October 18, 2022, https://electrek.co/2022/10/18/us-electric-vehicle-sales-by-maker-and-ev-model-through-q3-2022/.
26. Thea Riofrancos, "Electric Vehicles Alone Won't Take Us to a Decarbonized Future," The Hill, February 17, 2023, https://thehill.com/opinion/energy-environment/3861795-electric-vehicles-alone-wont-take-us-to-a-decarbonized-future/. See also Benjamin K. Sovacool et al., "Imagining Sustainable Energy and Mobility Transitions: Valence, Temporality, and Radicalism in 38 Visions of a Low-Carbon Future:," *Social Studies of Science* 50, no. 4 (2020): 642–79, https://doi.org/10.1177/0306312720915283; Jonny Lieberman, "You're Being Lied to About Electric Cars," MotorTrend, December 8, 2022, https://www.motortrend.com/features/truth-about-electric-cars-ad-why-you-are-being-lied-to/.

27. John Thøgersen et al., "Why Do People Continue Driving Conventional Cars Sespite Climate Change? Social-Psychological and Institutional Insights from a Survey of Norwegian Commuters," *Energy Research & Social Science* 79 (September 1, 2021): 102168, https://doi.org/10.1016/j.erss.2021.102168; Marianne Aasen et al., "The Limited Influence of Climate Norms on Leisure Air Travel," *Journal of Sustainable Tourism* (August 9, 2022): 1–19, https://doi.org/10.1080/09669582.2022.2097687; Arild Vatn et al., "What Role Do Climate Considerations Play in Consumption of Red Meat in Norway?," *Global Environmental Change* 73 (March 1, 2022): 102490, https://doi.org/10.1016/j.gloenvcha.2022.102490.
28. Naomi Oreskes and Erik M. Conway, *Merchants of Doubt: How a Handful of Scientists Obscured the Truth on Issues from Tobacco Smoke to Global Warming* (New York: Bloomsbury Publishing USA, 2011); Geoffrey Supran and Naomi Oreskes, "Assessing ExxonMobil's Climate Change Communications (1977–2014)," *Environmental Research Letters* 12, no. 8 (2017): 084019, https://doi.org/10.1088/1748-9326/aa815f; G. Supran, S. Rahmstorf, and N. Oreskes, "Assessing ExxonMobil's Global Warming Projections," *Science* 379, no. 6628 (January 13, 2023): eabk0063, https://doi.org/10.1126/science.abk0063.
29. Supran, Rahmstorf, and Oreskes, "Assessing ExxonMobil's Global Warming Projections." See also Oreskes and Conway, *Merchants of Doubt.*
30. Mark Kaufman, "The Devious Fossil Fuel Propaganda We All Use," Mashable, July 13, 2020, https://mashable.com/feature/carbon-footprint-pr-campaign-sham. See also Amy Westervelt, "Big Oil Is Trying to Make Climate Change Your Problem to Solve: Don't Let Them," *Rolling Stone* (blog), May 14, 2021, https://www.rollingstone.com/politics/politics-news/climate-change-exxonmobil-harvard-study-1169682/.
31. David Chandler, "Leaving Our Mark," *MIT News*, April 16, 2008, https://news.mit.edu/2008/footprint-tt0416.
32. Intergovernmental Panel on Climate Change (IPCC), "Chapter 4: Strengthening and Implementing the Global Response," in *Global Warming of 1.5°C. An IPCC Special Report on the Impacts of Global Warming of 1.5°C above Pre-Industrial Levels and Related Global Greenhouse Gas Emission Pathways, in the Context of Strengthening the Global Response to the Threat of Climate Change, Sustainable Development, and Efforts to Eradicate Poverty* (New York: Cambridge University Press, 2018), https://www.ipcc.ch/sr15/chapter/chapter-4/.

33. United Nations Environment Programme (UNEP), *Emissions Gap Report 2022* (Nairobi, Kenya: UNEP, October 21, 2022), http://www.unep.org/resources/emissions-gap-report-2022.
34. Naomi Oreskes, "Why Didn't They Act?," in *The Climate Book*, ed. Greta Thunberg (London: Penguin Random House, 2022), 29–31. See also Naomi Oreskes and Erik M. Conway, *The Big Myth: How American Business Taught Us to Loathe Government and Love the Free Market* (Bloomsbury Publishing, 2023).
35. Naomi Oreskes and Erik Conway, "The True Cost of the 'Free' Market Was Exposed by the Pandemic and Climate Change," *Time*, February 28, 2023, https://time.com/6258540/true-cost-of-the-free-market/. See also Oreskes and Conway, *The Big Myth*.
36. Toru Muta and Musa Erdogan, "The Global Energy Crisis Pushed Fossil Fuel Consumption Subsidies to an All-Time High in 2022," International Energy Agency (IEA), February 16, 2023, https://www.iea.org/commentaries/the-global-energy-crisis-pushed-fossil-fuel-consumption-subsidies-to-an-all-time-high-in-2022.
37. Helmut Anheier and Nuno Themudo, "Organizational Forms of Global Civil Society: Implications of Going Global," in *Global Civil Society 2002*, ed. M. Glasius, M. Kaldor, and H. K. Anheier (Oxford: Oxford University Press, 2002), 27. Dana R. Fisher and Jessica F. Green, "Understanding Disenfranchisement: Civil Society and Developing Countries' Influence and Participation in Global Governance for Sustainable Development," *Global Environmental Politics* 4, no. 3 (2004): 65–84. For a more general discussion, see William A. Gamson, *The Strategy of Social Protest* (Homewood, IL: Irwin-Dorsey, 1975); Kenneth T. Andrews, "How Protest Works," *New York Times*, October 21, 2017, sec. Opinion, https://www.nytimes.com/2017/10/21/opinion/sunday/how-protest-works.html.
38. W. Lance Bennett, *The Logic of Connective Action: Digital Media and the Personalization of Contentious Politics* (Cambridge: Cambridge University Press, 2013); Yannis Theocharis, Silia Vitoratou, and Javier Sajuria, "Civil Society in Times of Crisis: Understanding Collective Action Dynamics in Digitally-Enabled Volunteer Networks," *Journal of Computer-Mediated Communication* 22, no. 5 (2017): 248–65, https://doi.org/10.1111/jcc4.12194.
39. Chrys Salt and Jim Layzell, *Here We Go!: Women's Memories of the 1984/85 Miners Strike* (London Political Committee, Co-operative Retail Services

Limited, 1985); Elizabeth Cherry, "Veganism as a Cultural Movement: A Relational Approach," *Social Movement Studies* 5, no. 2 (September 1, 2006): 155–170, https://doi.org/10.1080/14742830600807543; Avelie Stuart et al., "'We May Be Pirates, but We Are Not Protesters': Identity in the Sea Shepherd Conservation Society," *Political Psychology* 34, no. 5 (2013): 753–777, https://www.jstor.org/stable/43783734; Christina Ergas, "A Model of Sustainable Living: Collective Identity in an Urban Ecovillage," *Organization & Environment* 23, no. 1 (March 1, 2010): 32–54, https://doi.org/10.1177/1086026609360324; Lucie Middlemiss, "The Effects of Community-Based Action for Sustainability on Participants' Lifestyles," *Local Environment* 16, no. 3 (March 1, 2011): 265–80, https://doi.org/10.1080/13549839.2011.566850; Ross Haenfler, Brett Johnson, and Ellis Jones, "Lifestyle Movements: Exploring the Intersection of Lifestyle and Social Movements," *Social Movement Studies* 11, no. 1 (January 1, 2012): 1–20, https://doi.org/10.1080/14742837.2012.640535; James M. Cronin, Mary B. McCarthy, and Alan M. Collins, "Covert Distinction: How Hipsters Practice Food-Based Resistance Strategies in the Production of Identity," *Consumption Markets & Culture* 17, no. 1 (January 2, 2014): 2–28, https://doi.org/10.1080/10253866.2012.678785; Clare Saunders et al., "Beyond the Activist Ghetto: A Deductive Blockmodelling Approach to Understanding the Relationship Between Contact with Environmental Organisations and Public Attitudes and Behaviour," *Social Movement Studies* 13, no. 1 (January 2, 2014): 158–77, https://doi.org/10.1080/14742837.2013.832623; Milena Büchs et al., "Identifying and Explaining Framing Strategies of Low Carbon Lifestyle Movement Organisations," *Global Environmental Change* 35 (November 1, 2015): 307–15, https://doi.org/10.1016/j.gloenvcha.2015.09.009; Wynes et al., "Measuring What Works;" Wynes and Nicholas, "The Climate Mitigation Gap."

40. Iljana Schubert, Judith I. M. de Groot, and Adrian C. Newton, "Challenging the Status Quo Through Social Influence: Changes in Sustainable Consumption Through the Influence of Social Networks," *Sustainability* 13, no. 10 (January 2021): 5513, https://doi.org/10.3390/su13105513.
41. Intergovernmental Panel on Climate Change Working Group 3, *Climate Change 2022: Mitigation of Climate Change*.
42. David S. Meyer and Sidney Tarrow, eds., *The Social Movement Society* (Lanham, MD: Rowman & Littlefield, 1997).
43. Isabella Kaminski, "Why 2023 Will Be a Watershed Year for Climate Litigation," *The Guardian*, January 4, 2023, https://www.theguardian

.com/environment/2023/jan/04/why-2023-will-be-a-watershed-year-for-climate-litigation; UN Environment Programme (UNEP), "Global Climate Litigation Report: 2023 Status Review," UNEP, 2023, https://doi.org/10.59117/20.500.11822/43008.

44. For more discussion of these tactics and their effects, see Dana R. Fisher and Sohana Nasrin, "Climate Activism and Its Effects," *Wiley Interdisciplinary Reviews: Climate Change* 12, no. 1 (2021): e683, https://doi.org/10.1002/wcc.683.
45. Dana R. Fisher, "Youth Political Participation: Bridging Activism and Electoral Politics," *Annual Review of Sociology* 38 (2012): 119–37; Dana R. Fisher, *American Resistance: From the Women's March to the Blue Wave* (New York: Columbia University Press, 2019).
46. Matthew H. Goldberg et al., "Oil and Gas Companies Invest in Legislators That Vote Against the Environment," *Proceedings of the National Academy of Sciences* 117, no. 10 (February 19, 2020): 5111–12, https://doi.org/10.1073/pnas.1922175117.
47. "Tell Our Leaders: Take the No Fossil Fuel Money Pledge," accessed February 24, 2023, https://nofossilfuelmoney.org/.
48. Brady Dennis and Dino Grandoni, "How Joe Biden's Surprisingly Ambitious Climate Plan Came Together," *Washington Post*, August 2, 2020, https://www.washingtonpost.com/climate-environment/how-joe-bidens-surprisingly-ambitious-climate-plan-came-together/2020/07/31/b73e78d0-cd11-11ea-91f1-28aca4d833a0_story.html.
49. Dana R. Fisher, "Youth Climate Activists Once Opposed Joe Biden. Now, They Say They'll Vote for Him," Politico, May 8, 2020, https://www.politico.com/news/magazine/2020/05/08/youth-climate-activists-joe-biden-survey-241068.
50. Evan Mceldowney, "Sunrise Movement Celebrates Major Climate Victories After Biden Takes Reins on Climate Action, Asserts Democrats Must Now Abolish Filibuster," January 27, 2021, Sunrise Movement, accessed February 24, 2023, https://www.sunrisemovement.org/press-releases/biden-climate-action/.
51. For an overview, see Dana R. Fisher, "Civil Society Protest and Participation: Civic Engagement Within the Multilateral Governance Regime," in *Emerging Forces in Environmental Governance*, ed. Norichika Kanie and Peter M. Haas (Tokyo: United Nations University Press, 2004), 176–99.
52. For more details, see Fisher, *American Resistance*.

53. For more detail, see Dana R Fisher, *National Governance and the Global Climate Change Regime* (Lanham, MD: Rowman & Littlefield, 2004).
54. Dana R. Fisher et al., "How Do Organizations Matter? Mobilization and Support for Participants at Five Globalization Protests," *Social Problems* 52, no. 1 (February 1, 2005): 108, https://doi.org/10.1525/sp.2005.52.1.102.
55. For a full discussion, see *Dana R. Fisher, "COP-15 in Copenhagen: How the Merging of Movements Left Civil Society out in the Cold," Global Environmental Politics 10, no. 2 (2010): 11–17.*
56. "With Marches Banned, Shoes Carry a Message," *New York Times*, November 29, 2015, silent-rally-activists-shoes.
57. *Global Climate March Report*, accessed December 31, 2022, https://350.org /global-climate-march/.
58. Graham St John, "Protestival: Global Days of Action and Carnivalized Politics in the Present," *Social Movement Studies* 7, no. 2 (2008): 167–90. See also Fisher, *American Resistance.*
59. Dana R. Fisher and Marije Boekkooi, "Mobilizing Friends and Strangers," *Information, Communication & Society* 13, no. 2 (March 1, 2010): 193–208, https://doi.org/10.1080/13691180902878385; 350.org, "Our History," accessed December 31, 2022, https://350.org/our-history/.
60. Fisher, interview with Bill McKibben, May 11, 2023.
61. Fisher, *American Resistance.*
62. Dana R. Fisher, Dawn Marie Dow, and Rashawn Ray, "The Demographics of the #resistance," The Conversation, June 1, 2017, http:// theconversation.com/the-demographics-of-the-resistance-77292.
63. Fisher and Nasrin, "Climate Activism and Its Effects."
64. Fisher, interview with Aaron Huertas, September 26, 2022.
65. Fisher, interview with Phil Aroneanu, August 16, 2022.
66. For an account of the unintended consequences of mobilization through these weaker, electronically mediated ties, see Zeynep Tufekci, *Twitter and Tear Gas: The Power and Fragility of Networked Protest* (New Haven, CT: Yale University Press, 2017). See also Fisher, *American Resistance*; Maurice Mitchell, "Building Resilient Organizations: Toward Joy and Durable Power in a Time of Crisis," *Convergence Magazine*, November 29, 2022, https://convergencemag.com/articles/building-resilient-organizations -toward-joy-and-durable-power-in-a-time-of-crisis/.
67. Fisher, *American Resistance.*
68. Dana R. Fisher, "Climate of Resistance: How the Climate Movement Connects to the Resistance," in *The Resistance: The Dawn of the*

Anti-Trump Opposition Movement, ed. David S. Meyer and Sidney Tarrow (New York: Oxford University Press, 2018), 109–26.

69. For details, see Fisher, "Climate of Resistance."
70. For an overview of the methodology used to collect these data, see Lorien Jasny and Dana R. Fisher, "How Networks of Social Movement Issues Motivate Climate Resistance," *Social Networks* (February 11, 2022), https://doi.org/10.1016/j.socnet.2022.02.002.
71. Fisher, *American Resistance.*
72. United States Census Bureau, "Census Bureau Releases New Educational Attainment Data," Press Release Number CB22-TPS.02, February 24, 2022, https://www.census.gov/newsroom/press-releases/2022/educational-attainment.html.
73. Marco Giugni and Maria T. Grasso, "Environmental Movements in Advanced Industrial Democracies: Heterogeneity, Transformation, and Institutionalization," *Annual Review of Environment and Resources* 40, no. 1 (2015): 337–61, https://doi.org/10.1146/annurev-environ-102014-021327; Leon Pieters et al., "Who Is Setting the Pace for Personal Sustainability?," Deloitte Insights (Deloitte Center for Integrated Research, April 2022), https://www2.deloitte.com/us/en/insights/topics/strategy/sustainable-action-climate-change.html.
74. J. Marlon et al., "Younger Americans Are Growing More Worried About Global Warming," Yale Program on Climate Change Communication (blog), December 15, 2022, https://climatecommunication.yale.edu/publications/younger-americans-are-growing-more-worried-about-global-warming/.
75. Marlon et al.
76. Dana R. Fisher, "The Youth Climate Summit Starts July 12: It's Full of Young Activists Trained in the Anti-Trump Movement.," *Washington Post*, July 12, 2019, sec. Monkey Cage, https://www.washingtonpost.com/politics/2019/07/12/youth-climate-summit-starts-today-its-full-young-activists-trained-anti-trump-movement/.
77. Damian Carrington, "School Climate Strikes: 1.4 Million People Took Part, Say Campaigners," *The Guardian*, March 19, 2019, sec. Environment, https://www.theguardian.com/environment/2019/mar/19/school-climate-strikes-more-than-1-million-took-part-say-campaigners-greta-thunberg.
78. 350.org Team, "7.6 Million People Demand Action After Week of Climate Strikes," September 28, 2019, https://350.org/7-million-people-demand-action-after-week-of-climate-strikes/.

79. For more details of my findings, see Dana R. Fisher and Sohana Nasrin, "Shifting Coalitions Within the Youth Climate Movement in the US," *Politics and Governance* 9, no. 2 (April 28, 2021): 112–23, https://doi.org/10.17645/pag.v9i2.3801; Sohana Nasrin and Dana R. Fisher, "Understanding Collective Identity in Virtual Spaces: A Study of the Youth Climate Movement," *American Behavioral Scientist* 66, no. 9 (2022): 1286–1308.
80. Fisher and Nasrin, "Shifting Coalitions Within the Youth Climate Movement."
81. These demographics are also similar to participants in environmental movements in the developed world. See Giugni and Grasso, "Environmental Movements in Advanced Industrial Democracies."
82. Dana R. Fisher, Lorien Jasny, and Dawn M. Dow, "Why Are We Here? Patterns of Intersectional Motivations Across the Resistance," *Mobilization: An International Quarterly* 23, no. 4 (December 1, 2018): 451–68, https://doi.org/10.17813/1086-671X-23-4-451.
83. For more discussion, see Fisher, *American Resistance.*
84. Lorien Jasny and Dana R. Fisher, "How Networks of Social Movement Issues Motivate Climate Resistance," *Social Networks* (February 11, 2022), https://doi.org/10.1016/j.socnet.2022.02.002.
85. Michael McCloskey, "Twenty Years of Change in the Environmental Movement: An Insider's View," *Society & Natural Resources* 4, no. 3 (July 1, 1991): 273–84, https://doi.org/10.1080/08941929109380760.
86. Liam Downey, "Environmental Injustice: Is Race or Income a Better Predictor?," *Social Science Quarterly* 79, no. 4 (1998): 766–78, https://www.jstor.org/stable/42863846; Andrew Jorgenson et al., "Power, Proximity, and Physiology: Does Income Inequality and Racial Composition Amplify the Impacts of Air Pollution on Life Expectancy in the United States?," *Environmental Research Letters* 15, no. 2 (2020), https://doi.org/10.1088/1748-9326/ab6789. See also Robert D. Bullard, *Dumping in Dixie: Race, Class, And Environmental Quality*, 3rd ed. (New York: Routledge, 1990); Robert D. Bullard and Beverly H. Wright, "The Quest for Environmental Equity: Mobilizing the African-American Community for Social Change," in *American Environmentalism: The US Environmental Movement, 1970–1990*, ed. Riley E. Dunlap and Angela G. Mertig (New York: Taylor & Francis, 1992).
87. For more details, see Jasny and Fisher, "How Networks of Social Movement Issues."

3. SAVING OURSELVES INVOLVES TAKING POWER

88. Arthur M. Schlesinger, "Biography of a Nation of Joiners," *The American Historical Review* 50, no. 1 (1944): 1–25, https://doi.org/10.2307/1843565.
89. Aldon Morris, "From Civil Rights to Black Lives Matter," *Scientific American*, February 3, 2021, https://www.scientificamerican.com/article/from-civil-rights-to-black-lives-matter1/. See also Oscar Berglund, "Disruptive Protest, Civil Disobedience & Direct Action," *Politics* (2023), https://journals.sagepub.com/doi/full/10.1177/02633957231176999; David Graeber, *Direct Action: An Ethnography* (Oakland, CA: AK Press, 2009).
90. Fisher, *American Resistance.*
91. Pieters et al., "Who Is Setting the Pace for Personal Sustainability?"

4. SAVING OURSELVES WON'T BE POPULAR AND IT WILL BE DISRUPTIVE

1. Loveday Morris, "Police Stuck in Mud, Greta Thunberg Detained at Coal Protest in Germany," *Washington Post*, January 17, 2023, https://www.washingtonpost.com/world/2023/01/17/germany-coal-village-mud/. See also International Energy Agency (IEA), "Coal," accessed March 5, 2023, https://www.iea.org/fuels-and-technologies/coal.
2. Loveday Morris, "Germany Portrays Itself as a Climate Leader, but It's Still Razing Villages for Coal Mines.," *Washington Post*, October 23, 2021, https://www.washingtonpost.com/world/2021/10/23/germany-coal-climate-cop26/. See also Katya Golubkova, "G7 Alarms Climate Activists over Support for Gas Investments," Reuters, May 20, 2023, https://www.reuters.com/business/energy/g7-brings-gas-investments-back-temporary-solution-dismay-climate-activists-2023-05-20/.
3. Transcribed from full speech by Greta Thunberg, "The Climate Event," Southbank Centre, October 30, 2022, https://www.youtube.com/watch?app=desktop&v=ropBOwPvmLM.
4. Transcribed from comments made by Al Gore, "We Are Still Failing Badly," Forbes Breaking News, January 18, 2023, https://www.youtube.com/watch?v=4-br-n9xTOc.
5. Just Transition Research Cooperative, "Mapping Just Transition(s) to a Low-Carbon World," UN Research Institute for Sustainable Development, December 2018, www.unrisd.org/jtrc-report2018.

6. Lorien Jasny and Dana R. Fisher, "How Networks of Social Movement Issues Motivate Climate Resistance," *Social Networks* (February 11, 2022), https://doi.org/10.1016/j.socnet.2022.02.002.
7. See The White House Briefing Room, "What Are They Saying: President Biden Takes Action to Build Healthy Communities and Advance Environmental Justice," April 22, 2023, https://www.whitehouse.gov/briefing-room/statements-releases/2023/04/22/what-they-are-saying-president-biden-takes-action-to-build-healthy-communities-and-advance-environmental-justice/.
8. Climate Justice Alliance, "The Inflation Reduction Act Is *Not* a Climate Justice Bill," press release, August 6, 2022, https://climatejusticealliance.org/the-inflation-reduction-act-is-not-a-climate-justice-bill/. See also Kate Aronoff, "The Manchin Climate Deal Is Both a Big Win and a Deal With the Devil," *The New Republic*, August 3, 2022, https://newrepublic.com/article/167272/manchin-climate-deal-big-win-deal-devil; Zack Colman, "'We've Been Sold Out': Enviro Justice Advocates Slam Biden's Climate Compromise," Politico, August 18, 2022, https://www.politico.com/news/2022/08/18/weve-been-sold-out-enviro-justice-advocates-slam-bidens-climate-compromise-00052048; Sarah Kaplan, "A Victory at Whose Expense? Climate Activists Grapple with Political Compromise," *Washington Post*, August 11, 2022, https://www.washingtonpost.com/climate-environment/2022/08/10/victory-whose-expense-climate-activists-grapple-with-political-compromise/.
9. Anthony Rogers-Wright, "Why the Inflation Reduction Act Is Less a 'Climate Bill' and More a Poison Pill for Black and Indigenous Communities and Movements," Black Agenda Report, August 24, 2022, http://www.blackagendareport.com/why-inflation-reduction-act-less-climate-bill-and-more-poison-pill-black-and-indigenous-communities. See also Rhiana Gunn-Wright, "Our Green Transition May Leave Black People Behind," *Hammer & Hope* no. 2 (summer 2023), https://hammerandhope.org/article/climate-green-new-deal.
10. Dana R. Fisher, *American Resistance: From the Women's March to the Blue Wave* (New York: Columbia University Press, 2019).
11. This list is not meant to be exhaustive. For details, see Fisher, *American Resistance*.
12. Mary Jordan and Scott Clement, "Echoes of Vietnam: Millions of Americans Are Taking to the Streets," *Washington Post*, April 6, 2018, https://www.washingtonpost.com/news/national/wp/2018/04/06/feature

/in-reaction-to-trump-millions-of-americans-are-joining-protests-and-getting-political/. For evidence of the growing urge to protest, see R. J. Reinhart, "One in Three Americans Have Felt Urge to Protest," Gallup News, August 24, 2018, https://news.gallup.com/poll/241634/one-three-americans-felt-urge-protest.aspx?.

13. For more details, see Dana R. Fisher, "Climate of Resistance: How the Climate Movement Connects to the Resistance," in *The Resistance: The Dawn of the Anti-Trump Opposition Movement*, ed. David S. Meyer and Sidney Tarrow (New York: Oxford University Press, 2018), 109–26.
14. Fisher, *American Resistance*.
15. Sanya Mansoor, "93 Percent of Black Lives Matter Protests Have Been Peaceful, New Report Finds," *Time*, September 5, 2020, https://time.com/5886348/report-peaceful-protests/.
16. See, for example, Michelle Goldberg, "1, 2, 3, 4, Trump Can't Rule Us Anymore," *New York Times*, October 22, 2019, sec. Opinion, https://www.nytimes.com/2019/10/21/opinion/trump-protests.html; David Leonhardt, "Want Trump to Go? Take to the Streets," *New York Times*, October 20, 2019, sec. Opinion, https://www.nytimes.com/2019/10/20/opinion/trump-impeachment-protests.html; Matthew Yglesias, "Impeachment Is Too Important to Leave to Congress—It's Going to Take Mass Mobilization," Vox, October 18, 2019, https://www.vox.com/policy-and-politics/2019/10/18/20905686/resistance-protest-impeachment-rallies-trump.
17. Dana R Fisher, "The Original Women's Marchers Are Still a Political Force," *Washington Post*, November 3, 2020, https://www.washingtonpost.com/politics/2020/11/03/original-womens-marchers-are-still-political-force/.
18. For an overview, see Rachel Kleinfeld, "The Rise of Political Violence in the United States," *Journal of Democracy* 32, no. 4 (October 2021): 160–76, https://www.journalofdemocracy.org/articles/the-rise-of-political-violence-in-the-united-states/.
19. Michael T. Heaney and Fabio Rojas, *Party in the Street: The Antiwar Movement and the Democratic Party after 9/11* (New York: Cambridge University Press, 2015). See also Micah L. Sifry, "Obama's Lost Army," *New Republic*, February 9, 2017, https://newrepublic.com/article/140245/obamas-lost-army-inside-fall-grassroots-machine.
20. Dana R Fisher, "Welcome to the Graveyard of the Progressive Movement," American Resistance (blog), January 20, 2022, https://american

resistancebook.com/2022/01/20/welcome-to-the-graveyard-of-the-progressive-movement/.

21. Fisher, off-the-record interview with climate leader, summer 2022.
22. On professionalization, see Suzanne Staggenborg, "The Consequences of Professionalization and Formalization in the Pro-Choice Movement," *American Sociological Review* 53, no. 4 (1988): 585–605, https://doi.org/10.2307/2095851.
23. Ben Adler, "Biden Granted More Oil and Gas Drilling Permits than Trump in His First 2 Years in Office," Yahoo! News, January 25, 2023, https://news.yahoo.com/biden-granted-more-oil-and-gas-drilling-permits-than-trump-in-his-first-2-years-in-office-190528616.html.
24. See particularly Gunn-Wright, "Our Green Transition May Leave Black People Behind."
25. Herbert H. Haines, "Black Radicalization and the Funding of Civil Rights: 1957–1970," *Social Problems* 32, no. 1 (October 1, 1984): 31–43, https://doi.org/10.2307/800260; Lewis M. Killian, "The Significance of Extremism in the Black Revolution," *Social Problems* 20, no. 1 (1972): 41–49. See also Doug McAdam, *Political Process and the Development of Black Insurgency, 1930–1970*, 2nd ed. (Chicago: University of Chicago Press, 1982); Aldon D. Morris, *The Origins of the Civil Rights Movement: Black Communities Organizing for Change* (New York: Free Press, 1986); Joshua Bloom and Waldo E. Martin Jr, *Black Against Empire: The History and Politics of the Black Panther Party, with a New Preface*, 2016.
26. Myra Marx Ferree and Beth B. Hess, *Controversy and Coalition: The New Feminist Movement Across Four Decades of Change*, 3rd ed. (New York: Routledge, 2000); Holly J. McCammon, Erin M. Bergner, and Sandra C. Arch, "'Are You One of Those Women?' Within-Movement Conflict, Radical Flank Effects, and Social Movement Political Outcomes," *Mobilization: An International Quarterly* 20, no. 2 (June 1, 2015): 157–78, https://doi.org/10.17813/1086-671X-20-2-157.
27. Amin Ghaziani, *The Dividends of Dissent: How Conflict and Culture Work in Lesbian and Gay Marches on Washington*, illustrated ed. (Chicago: University of Chicago Press, 2008); Amin Ghaziani and Delia Baldassarri, "Cultural Anchors and the Organization of Differences: A Multi-Method Analysis of LGBT Marches on Washington," *American Sociological Review* 76, no. 2 (April 1, 2011): 179–206, https://doi.org/10.1177/0003122411401252.

28. See in particular McCammon, Bergner, and Arch, "'Are You One of Those Women?'"
29. Brent Simpson, Robb Willer, and Matthew Feinberg, "Radical Flanks of Social Movements Can Increase Support for Moderate Factions," *PNAS Nexus* 1, no. 3 (July 1, 2022): 1, https://doi.org/10.1093/pnasnexus/pgac110. For an overview, see Haines, "Black Radicalization and the Funding of Civil Rights." Also see Dana R. Fisher, "Understanding the Growing Radical Flank of the Climate Movement as the World Burns," Brookings Institution (2023), accessed September 2, 2023, https://www.brookings.edu/articles/understanding-the-growing-radical-flank-of-the-climate-movement-as-the-world-burns/.
30. See McAdam, *Political Process and the Development of Black Insurgency*, for a full discussion.
31. See, for example, Stokley Carmichael and Charles V. Hamilton, *Black Power: Politics of Liberation in America* (New York: Random House, 1967), https://books.google.com/books/about/Black_Power.html?id=ysI_HYSKT9kC; McAdam, *Political Process and the Development of Black Insurgency*; Morris, *The Origins of the Civil Rights Movement*; Bloom and Martin, *Black Against Empire*; Joshua Bloom, "The Dynamics of Repression and Insurgent Practice in the Black Liberation Struggle," *American Journal of Sociology* 126, no. 2 (September 1, 2020): 195–259, https://doi.org/10.1086/711672.
32. Fisher, interview with Bill McKibben May 11, 2023.
33. Todd Schifeling and Andrew J. Hoffman, "Bill McKibben's Influence on U.S. Climate Change Discourse: Shifting Field-Level Debates Through Radical Flank Effects," *Organization & Environment* 32, no. 3 (September 1, 2019): 213–33, https://doi.org/10.1177/1086026617744278.
34. United States Climate Action Network (USCAN), "Who We Are," accessed May 22, 2023, https://www.usclimatenetwork.org/who-we-are.
35. Fisher, interview with Keya Chatterjee, August 11, 2022. Chatterjee stepped down as executive director of USCAN in February 2023: Stephanie Ready, "Farewell and Thank You," USCAN, November 2, 2022, https://www.usclimatenetwork.org/farewell_and_thank_you_keya.
36. Ruairí Arrieta-Kenna, "The Sunrise Movement Actually Changed the Democratic Conversation: So What Do You Do for a Sequel?," Politico Magazine, June 16, 2019, https://politi.co/2WJnIa2.
37. Email to Dana Fisher from Varshini Prakash, April 6, 2023.
38. "We Are Sunrise Movement: We Are the Climate Revolution," Sunrise Movement, accessed May 22, 2023, https://www.sunrisemovement.org/.

39. Fisher, off-the-record interview with climate leader, summer 2022.
40. It is worth noting that many planned protests in Washington, DC, were canceled after the insurrection at the U.S. Capitol on January 6, 2021, and the threats of violence during the inauguration of Joseph Biden, including protests that had been scheduled to coincide with the inauguration. See Susannah Cullinane and Ray Sanchez, "'Armed Protests' Warning Puts Officials on Alert This Weekend Ahead of Biden Inauguration," CNN, updated January 17, 2021, https://www.cnn.com/2021/01/16/us/inauguration-protests-saturday/index.html. See also Fisher, "Welcome to the Graveyard of the Progressive Movement."
41. For more details, see Fisher, "Understanding the Growing Radical Flank of the Climate Movement as the World Burns."
42. Extinction Rebellion, "This Is an Emergency," accessed September 4, 2023, https://rebellion.global/.
43. See Extinction Rebellion Global, "There has long been a tension between incrementalists in the environmental movement, and people like us who demand radical change . . . ," November 18, 2021, https://twitter.com/ExtinctionR/status/1461278569393496068?t.
44. Fisher, interview with Yi Mun, July 25, 2022.
45. Fisher, off-the-record interview with climate leader, summer 2022.
46. Vanessa Williamson and Dana R. Fisher, "It's Time for Democrats to Stop 'Clapping for Tinkerbell,'" *The Nation*, June 10, 2022, https://www.thenation.com/article/politics/democrats-clapping-tinkerbell/.
47. Catie Edmondson, "The Meaning of the Jan. 6 Gallows Erected in Front of the Capitol." *New York Times*, June 16, 2022, https://www.nytimes.com/2022/06/16/us/politics/jan-6-gallows.html.
48. See, for example, Jonathan Weisman and Reid J. Epstein, "G.O.P. Declares Jan. 6 Attack 'Legitimate Political Discourse,'" *New York Times*, February 4, 2022, https://www.nytimes.com/2022/02/04/us/politics/republicans-jan-6-cheney-censure.html.
49. Julia Jacobo, "Alaska Oil Drilling Willow Project Approved, Despite Viral Protests," ABC News, March 14, 2023, https://abcnews.go.com/US/willow-project-confirmed-despite-protests-experts-explain-young/story?id=97742724.
50. Gen-Z for Change, "We are Gen-Z for Change, a coalition of Gen-Z activists who have worked with @POTUS for the duration of his presidency . . . ," March 13, 2023, https://twitter.com/genzforchange/status/1635449569273159683.

51. John Wagner and Mariana Alfaro, "'This Is Your Victory,' Biden Celebrates Inflation Reduction Act," *Washington Post*, September 13, 2022, https://www.washingtonpost.com/politics/2022/09/13/biden-inflation-reduction-act/.
52. As quoted in Lexi McManamin, "Willow Project Approved by Biden Despite Mass Protest by Gen Z, Climate Activists," Yahoo Life, March 13, 2023, https://www.yahoo.com/lifestyle/stopwillow-gen-z-organizes-millions-154218406.html.
53. Julia Conley, "Gen Z for Change Leader Interrupts Biden Press Secretary to Demand Climate Action," Common Dreams, July 28, 2023, https://www.commondreams.org/news/gen-z-protester-interrupts.
54. This list is not meant to be exhaustive. For an overview, see Oscar Berglund, "Disruptive Protest, Civil Disobedience & Direct Action," *Politics* (June 5, 2023), https://doi.org/10.1177/02633957231176999. See also David Graeber, *Direct Action: An Ethnography* (Oakland, CA: AK Press, 2009). For a historical overview, see Morris, *The Origins of the Civil Rights Movement*, chap. 3.
55. Shannon Osaka, "From Crashing 'The View' to Tomato Soup: Climate Protests Get Weird," *Washington Post*, October 24, 2022, https://www.washingtonpost.com/climate-environment/2022/10/14/tomato-soup-sunflowers-climate-protest/.
56. For details, see Climate Emergency Fund, "Disruptive Climate Campaigns in 11 Countries," accessed November 3, 2022, https://www.climateemergencyfund.org/the-plan.
57. For an overview, see Osaka, "From Crashing 'The View' to Tomato Soup."
58. Climate Emergency Fund, "About: We Fund the Movement's Leading Edge," accessed December 17, 2022, https://www.climateemergencyfund.org/about.
59. Samira Shackle, "'We'll Be Hated, but It Will Stir Things Up': Insulate Britain on What Happened Next—and Being Right All Along," *Guardian*, December 17, 2022, sec. Environment, https://www.theguardian.com/environment/2022/dec/17/insulate-britain-on-what-happened-next-energy-crisis.
60. Declare Emergency, "Frequently Asked Questions," accessed March 18, 2023, https://www.declareemergency.org/faq.
61. Fisher, interview with Nora Swisher, March 17, 2023.
62. Oscar Berglund and Daniel Schmidt, *Extinction Rebellion and Climate Change Activism: Breaking the Law to Change the World* (Cham: Springer International, 2020), 2, https://doi.org/10.1007/978-3-030-48359-3.

63. Shackle, "'We'll Be Hated, but It Will Stir Things Up.'" See also Omar Wasow, "Agenda Seeding: How 1960s Black Protests Moved Elites, Public Opinion and Voting," *American Political Science Review* 114, no. 3 (2020): 638–59, https://doi.org/10.1017/S000305542000009X.
64. Fire Drill Fridays, "History of Fire Drill Fridays," accessed March 21, 2023, https://firedrillfridays.org/history/.
65. Tyler Hersko, "Jane Fonda Is Building an Army to Defend the Earth with Fire Drill Fridays Movement," IndieWire (blog), February 8, 2020, https://www.indiewire.com/2020/02/jane-fonda-fire-drill-fridays-protests-arrests-los-angeles-1202209628/.
66. Fonda was arrested five times during this period. For details, see Hersko, "Jane Fonda Is Building an Army."
67. Scientist Rebellion, "Our Positions and Demands," accessed September 4, 2023, https://scientistrebellion.org/about-us/our-positions-and-demands/.
68. Rose Abramoff, "I'm a Scientist Who Spoke Up About Climate Change: My Employer Fired Me.," *New York Times*, January 10, 2023, sec. Opinion, https://www.nytimes.com/2023/01/10/opinion/scientist-fired-climate-change-activism.html.
69. Fisher, interview with Rose Abramoff, January 12, 2023.
70. Editorial Board, "Attacking Art Isn't Climate 'Protest': It's Vandalism.," *Washington Post*, May 8, 2023, https://www.washingtonpost.com/opinions/2023/05/08/climate-protest-degas-national-gallery/.
71. Colin Davis, "Just Stop Oil: Do Radical Protests Turn the Public Away from a Cause? Here's the Evidence," The Conversation, October 21, 2022, http://theconversation.com/just-stop-oil-do-radical-protests-turn-the-public-away-from-a-cause-heres-the-evidence-192901. See also Simpson, Willer, and Feinberg, "Radical Flanks of Social Movements;" James Ozden and Sam Glover, "The Radical Flank Effect of Just Stop Oil," Social Change Lab, December 2022, https://www.socialchangelab.org/_files/ugd/503ba4_a184ae5bbce24c228d07eda25566dc13.pdf. But see also Matthew Feinberg, Robb Willer, and Chloe Kovacheff, "The Activist's Dilemma: Extreme Protest Actions Reduce Popular Support for Social Movements," *Journal of Personality and Social Psychology* 119 (2020): 1086–1111, https://doi.org/10.1037/pspi0000230.
72. McCammon, Bergner, and Arch, "'Are You One of Those Women?'"
73. Doug McAdam, "Tactical Innovation and the Pace of Insurgency," *American Sociological Review* 48, no. 6 (1983): 752, https://doi.org/10.2307/2095322.

74. Anthony Leiserowitz et al., "Politics & Global Warming, April 2022," Yale Program on Climate Change Communication., April 2022, sec. 7, https://climatecommunication.yale.edu/publications/politics-global-warming-april-2022/.
75. Dana R. Fisher and Sohana Nasrin, "Climate Activism and Its Effects," *Wiley Interdisciplinary Reviews: Climate Change* 12, no. 100 (2021): e683, https://doi.org/10.1002/wcc.683.
76. Charles Tilly, "Speaking Your Mind Without Elections, Surveys, or Social Movements," *Public Opinion Quarterly* 47, no. 4 (January 1, 1983): 461–78, https://doi.org/10.1086/268805; Sidney Tarrow, "Cycles of Collective Action: Between Moments of Madness and the Repertoire of Contention," *Social Science History* 17, no. 2 (1993): 281–307, https://doi.org/10.2307/1171283; Sidney Tarrow, *Power in Movement: Social Movements and Contentious Politics* (Cambridge: Cambridge University Press, 1998); Charles Tilly, *Regimes and Repertoires* (Chicago: University of Chicago Press, 2010).
77. Email to Dana Fisher from Varshini Prakash, Executive Director of Sunrise Movement, April 6, 2023.
78. Emma Howard, "Harvard Divestment Campaigners Gear Up for a Week of Action," *Guardian*, April 13, 2015, sec. Environment, https://www.theguardian.com/environment/2015/apr/13/harvard-divestment-campaigners-gear-up-for-a-week-of-action.
79. Michelle G. Kruilla, "Hundreds of Divestment Protesters Storm Field, Interrupting Harvard-Yale Game," *Harvard Crimson*, November 24, 2019, https://www.thecrimson.com/article/2019/11/24/protesters-interrupt-the-game/.
80. Fisher, personal correspondence with Ilana Cohen, May 24, 2023.
81. Fossil Free Research, "Open Letter," accessed May 23, 2023, https://fossilfreeresearch.org/letter/.
82. Fisher, interview with Jacob Lowe, December 13, 2022.
83. Third Act, "Our Work," accessed March 21, 2023, https://thirdact.org/our-work/.
84. Oliver Milman, "'We Have Money and Power': Older Americans to Blockade Banks in Climate Protest," *Guardian*, March 19, 2023, sec. Environment, https://www.theguardian.com/environment/2023/mar/19/climate-crisis-protest-environment-third-act-bill-mckibben. Details also provided through author's direct correspondence with McKibben, May 2023.

85. Dayenu, "Our Mission," accessed January 4, 2023, https://dayenu.org/who-we-are/#our-mission.
86. Email correspondence to members received by Fisher, October 14, 2021.
87. Fisher, email correspondence with Jennie Rosenn, January 3, 2023.
88. Fisher, interview with Bill McKibben, May 11, 2023.
89. Fisher, email communication with Bill McKibben, May 25, 2023.
90. Haines, "Black Radicalization and the Funding of Civil Rights." See also Killian, "The Significance of Extremism;" Bloom and Martin, *Black Against Empire*.
91. Aldon Morris, "Black Southern Student Sit-in Movement: An Analysis of Internal Organization," *American Sociological Review* 46, no. 6 (1981): 748, https://doi.org/10.2307/2095077. See also Morris, *The Origins of the Civil Rights Movement*.
92. Fisher, off-the-record interview with climate organizer, summer 2022.
93. Fisher, off-the-record interview with climate organizer, summer 2022.
94. See, for example, Jared Gans, "Climate Protesters Face Federal Charges for Smearing Paint on Degas Sculpture Case," The Hill, May 27, 2023, https://thehill.com/regulation/court-battles/4023504-climate-protesters-face-federal-charges-for-smearing-paint-on-degas-sculpture-case/; Kristoffer Tigue, "Disruptive Climate Protests Spur Police Raids in Germany and the US," Inside Climate News (blog), June 2, 2023, https://insideclimatenews.org/news/02062023/todays-climate-police-raids-climate-protests/; Natasha Lennard and Akela Lacy, "Activists Face Felonies for Distributing Flyers on 'Cop City' Protester Killing," The Intercept, May 3, 2023, https://theintercept.com/2023/05/02/cop-city-activists-arrest-flyers/.
95. For an overview, see Christian Davenport, "State Repression and Political Order," *Annual Review of Political Science* 10, no. 1 (2007): 1–23, https://doi.org/10.1146/annurev.polisci.10.101405.143216.
96. See Rory McVeigh and Kevin Estep, *The Politics of Losing: Trump, the Klan, and the Mainstreaming of Resentment* (New York: Columbia University Press, 2019); Bloom and Martin, *Black Against Empire*; Bloom, "The Dynamics of Repression and Insurgent Practice." See also Kenneth T. Andrews, *Freedom Is a Constant Struggle: The Mississippi Civil Rights Movement and Its Legacy* (Chicago: University of Chicago Press, 2004), https://press.uchicago.edu/ucp/books/book/chicago/F/bo3618798.html.
97. McAdam, "Tactical Innovation and the Pace of Insurgency," 746. Other scholars have also noted how repression, in some cases, mobilized more

people to join the movement. Bloom, "The Dynamics of Repression and Insurgent Practice."

98. James M. Jasper and Jane D. Poulsen, "Recruiting Strangers and Friends: Moral Shocks and Social Networks in Animal Rights and Anti-Nuclear Protests," *Social Problems* 42, no. 4 (1995): 498, https://doi.org/10.2307/3097043. See also James M. Jasper, *The Art of Moral Protest: Culture, Biography, and Creativity in Social Movements* (Chicago: University of Chicago Press, 1997); Fisher, *American Resistance*.
99. Fisher, interview with Keya Chatterjee, August 11, 2022.
100. Fisher, interview with Bill McKibben, May 11, 2023.
101. Fisher, interview with Yi Mun, July 25, 2022.
102. Todd Gitlin, *The Whole World Is Watching: Mass Media in the Making and Unmaking of the New Left* (Berkeley: University of California Press, 2003).

5. SAVING OURSELVES WILL TAKE A DISASTER (OR MANY)

1. Elizabeth Weise and Doyle Rice, "As Canada Burns, Smoke Makes US Air Unhealthy and Skies Eerie: Is Climate Change to Blame?," *USA Today*, June 7, 2023, https://www.usatoday.com/story/news/nation/2023/06/07/see-wildfire-smoke-map-of-us-canada/70293572007/.
2. Justin Trudeau, "We're Seeing More and More of These Fires Because of Climate Change . . . ," Twitter, June 7, 2023, https://twitter.com/JustinTrudeau/status/1666586130571993088?t=v_wxnjD5Y1Mo_a2J6Ql9uQ&s=03.
3. NBC News, "New York City Has the Worst Air Quality in the World as Smoke from Canadian Wildfires Rolls In," updated June 7, 2023, https://www.nbcnews.com/news/us-news/live-blog/unhealthy-air-quality-canada-wildfires-live-updates-rcna88092.
4. Magan Carty, "World on Fire: 2023 Is Canada's Worst Wildfire Season on Record—and It's Not over Yet," CBC, September 4, 2023, https://www.cbc.ca/radio/ideas/world-on-fire-canada-s-worst-wildfire-season-on-record-1.6946472.
5. The White House, "The Potential Economic Impacts of Various Debt Ceiling Scenarios," (U.S. White House blog), May 3, 2023, https://www.whitehouse.gov/cea/written-materials/2023/05/03/debt-ceiling-scenarios/.

6. Coral Davenport and Brad Plumer, "Debt Deal Includes a Green Light for a Contentious Pipeline," *New York Times*, May 30, 2023, sec. Climate, https://www.nytimes.com/2023/05/30/climate/mountain-valley-pipe .html.
7. Vanessa Williamson and Dana R. Fisher, "It's Time for Democrats to Stop 'Clapping for Tinkerbell,'" The Nation, June 10, 2022, https://www .thenation.com/article/politics/democrats-clapping-tinkerbell/.
8. Lisa Friedman, "Climate Groups Back Biden Despite Broken Promises on Oil Drilling," *New York Times*, June 15, 2023, sec. Climate, https:// www.nytimes.com/2023/06/14/climate/environmental-groups-endorse -biden.html.
9. The U.S. Energy Information Agency (EIA) provided evidence of this strategy working and reported that U.S. oil production was the rise. See Arathy Somasekhar, "US 2023 Oil Output to Rise More than Previously Expected, EIA Says," Reuters, June 6, 2023, sec. Energy, https://www .reuters.com/business/energy/us-2023-oil-output-rise-more-than-previously -expected-eia-says-2023-06-06/.
10. Timothy Puko, "'Terrible Public Policy': Why the Debt Deal Infuriates Climate Activists," *Washington Post*, May 30, 2023, https://www.washington post.com/climate-environment/2023/05/29/debt-ceiling-deal-climate -pipeline/; Brian Dabbs, "Biden Fossil Fuel Boost Creates Political Storm on His Left," E&E News, May 26, 2023, https://www.eenews.net /articles/biden-fossil-fuel-boost-creates-political-storm-on-his-left/.
11. The administration was in the process of rolling out new rules to regulate emissions from power plants. However, preliminary assessments determine that the new regulations will be a boon for natural gas, and carbon capture and storage. See Rebecca Leber, "4 Winners and 1 Loser in the EPA's Historic Move to Limit Power Plant Pollution," Vox, May 13, 2023, https://www.vox.com/policy/2023/5/13/23719080/environmental -protection-agency-limit-power-plant-pollution.
12. Vera Eckert and Tom Sims, "Energy Crisis Fuels Coal Comeback in Germany," Reuters, December 16, 2022, https://www.reuters.com/markets /commodities/energy-crisis-fuels-coal-comeback-germany-2022-12-16/; Energy Mix, "Record Fossil Extraction from Canada, U.S., Norway Despite Fervent Climate Pledges," The Energy Mix (blog), February 2, 2022, https://www.theenergymix.com/2022/02/02/record-fossil-extraction -from-canada-u-s-norway-despite-fervent-climate-pledges/; Nerijus Adomaitis and Gwladys Fouce, "Norway Plans to Offer Record Number

of Arctic Oil, Gas Exploration Blocks," Reuters, January 24, 2023, https://www.reuters.com/business/energy/norway-offers-up-92-new-oil-gas-exploration-blocks-2023-01-24/.

13. The White House, "G7 Hiroshima Leaders' Communiqué," Briefing Room press release, May 20, 2023, https://www.whitehouse.gov/briefing-room/statements-releases/2023/05/20/g7-hiroshima-leaders-communique.
14. Energy Institute, *Statistical Review of World Energy 2023*, accessed September 5, 2023, https://www.energyinst.org/statistical-review/home; see also Reuters, "Renewables Growth Did Not Dent Fossil Fuel Dominance in 2022, Report Says," Reuters, June 26, 2023, https://www.reuters.com/business/energy/renewables-growth-did-not-dent-fossil-fuel-dominance-2022-statistical-review-2023-06-25/.
15. Fiona Harvey, "UK Missing Climate Targets on Nearly Every Front, Say Government's Advisers," *The Guardian*, June 27, 2023, https://www.theguardian.com/technology/2023/jun/28/uk-has-made-no-progress-on-climate-plan-say-governments-own-advisers. For information about the committee, see Climate Change Committee (CCC), "About Us," accessed September 4, 2023 https://www.theccc.org.uk/about/.
16. "Norway Approves More than $18 Billion in Oil, Gas Investments," Reuters, June 28, 2023, https://www.reuters.com/business/energy/norway-approves-more-than-18-bln-oil-gas-investments-2023-06-28/.
17. National Oceanic and Atmospheric Administration (NOAA), Global Monitoring Laboratory, "Trends in Atmospheric Carbon Dioxide," accessed June 6, 2023, https://gml.noaa.gov/ccgg/trends/.
18. Intergovernmental Panel on Climate Change (IPCC), *Global Warming of 1.5°C.* (Cambridge: Cambridge University Press, 2018), https://www.ipcc.ch/sr15/.
19. Piers M. Forster et al., "Indicators of Global Climate Change 2022: Annual Update of Large-Scale Indicators of the State of the Climate System and Human Influence," *Earth System Science Data* 15, no. 6 (June 8, 2023): 2295–2327, https://doi.org/10.5194/essd-15-2295-2023.
20. World Meteorological Organization (WMO), "Earth Had Hottest Three-Month Period on Record, with Unprecedented Sea Surface Temperatures and Much Extreme Weather," September 6, 2023, https://public.wmo.int/en/media/press-release/earth-had-hottest-three-month-period-record-unprecedented-sea-surface.
21. For details, see United Nations Treaty Collection, "Chapter XXVII: Environment," Paris Agreement, December 12, 2015, status as at May 8, 2023,

https://treaties.un.org/Pages/ViewDetails.aspx?src=TREATY&mtdsg_no=XXVII-7-d&chapter=27&clang=_en.

22. The full speech is available at United Nations, "Secretary-General's Press Conference—on Climate," June 15, 2023, https://www.un.org/sg/en/content/sg/press-encounter/2023-06-15/secretary-generals-press-conference-climate.
23. See, for example, Jared Gans, "Climate Protesters Face Federal Charges for Smearing Paint on Degas Sculpture Case," The Hill, May 27, 2023, https://thehill.com/regulation/court-battles/4023504-climate-protesters-face-federal-charges-for-smearing-paint-on-degas-sculpture-case/; Kristoffer Tigue, "Disruptive Climate Protests Spur Police Raids in Germany and the US," Inside Climate News (blog), June 2, 2023, https://insideclimatenews.org/news/02062023/todays-climate-police-raids-climate-protests/; Natasha Lennard and Akela Lacy, "Activists Face Felonies for Distributing Flyers on 'Cop City' Protester Killing," The Intercept, May 3, 2023, https://theintercept.com/2023/05/02/cop-city-activists-arrest-flyers/; Terry Macalister, "UK Climate Activists Held in Jail for up to Six Months Before Trial," *Guardian*, September 23, 2022, https://www.theguardian.com/environment/2022/sep/23/uk-climate-activists-held-in-jail-for-up-to-six-months-before-trial. See also Dana R. Fisher, "What History Tells Us About Protest in an Era of Vigilante Violence," Slate, November 24, 2021, https://slate.com/news-and-politics/2021/11/what-the-rittenhouse-verdict-means-for-the-future-of-protest-in-america.html.
24. Anthony Leiserowitz et al., "Climate Change in the American Mind: Beliefs and Attitudes, Spring 2023," Yale Program on Climate Change Communication (blog), June 8, 2023, https://climatecommunication.yale.edu/publications/climate-change-in-the-american-mind-beliefs-attitudes-spring-2023/.
25. Ellie Silverman, "Climate Advocates Protest Mountain Valley Pipeline Outside White House," *Washington Post*, June 8, 2023, https://www.washingtonpost.com/dc-md-va/2023/06/08/mountain-valley-pipeline-protest-white-house-debt-ceiling-appalachia/?s=03.
26. Will Steffen et al., "Trajectories of the Earth System in the Anthropocene," *Proceedings of the National Academy of Sciences* 115, no. 33 (August 1, 2018): 201810141, https://doi.org/10.1073/pnas.1810141115; Timothy M. Lenton et al., "Climate Tipping Points—Too Risky to Bet Against," *Nature* 575, no. 7784 (November 2019): 592–95, https://doi.org/10.1038

/d41586-019-03595-0. See also Simon Willcock et al., "Earlier Collapse of Anthropocene Ecosystems Driven by Multiple Faster and Noisier Drivers," *Nature Sustainability* (June 22, 2023): 1–12, https://doi.org/10.1038/s41893-023-01157-x.

27. Timothy M. Lenton et al., "Quantifying the Human Cost of Global Warming," *Nature Sustainability* (May 22, 2023): 1, https://doi.org/10.1038/s41893-023-01132-6.
28. Fisher, interview with Jonathan Pershing, February 11, 2022.
29. Kendra McSweeney and Oliver T. Coomes, "Climate-Related Disaster Opens a Window of Opportunity for Rural Poor in Northeastern Honduras," *Proceedings of the National Academy of Sciences* 108, no. 13 (March 29, 2011): 5203–8, https://doi.org/10.1073/pnas.1014123108.
30. Fisher, interview with Bill McKibben, May 11, 2023.
31. Fisher, off-the-record interview with climate leader, summer 2022.
32. Leon Pieters et al., "Who Is Setting the Pace for Personal Sustainability?," Deloitte Insights (Deloitte Center for Integrated Research, April 2022), https://www2.deloitte.com/us/en/insights/topics/strategy/sustainable-action-climate-change.html.
33. Pieters et al., 4–6.
34. Erica Chenoweth, "Questions, Answers, and Some Cautionary Updates Regarding the 3.5 Percent Rule," Carr Center Discussion Paper (Carr Center for Human Rights Policy, Harvard Kennedy School, April 2020), https://carrcenter.hks.harvard.edu/files/cchr/files/CCDP_005.pdf. See also Oscar Berglund and Daniel Schmidt, *Extinction Rebellion and Climate Change Activism: Breaking the Law to Change the World* (Cham: Springer International, 2020), https://doi.org/10.1007/978-3-030-48359-3.
35. Jürgen Habermas, *Legitimation Crisis* (Boston: Beacon Press, 1975).
36. See in particular Dana R. Fisher, *Activism, Inc.: How the Outsourcing of Grassroots Campaigns Is Strangling Progressive Politics in America* (Stanford, CA: Stanford University Press, 2006); Dana R. Fisher, *American Resistance: From the Women's March to the Blue Wave* (New York: Columbia University Press, 2019).
37. Aldon Morris, "Black Southern Student Sit-In Movement: An Analysis of Internal Organization," *American Sociological Review* 46, no. 6 (1981): 748, https://doi.org/10.2307/2095077.
38. Robert D. Putnam, *Bowling Alone: The Collapse and Revival of American Community* (New York: Touchstone, 2000); Carmen Sirianni and Lewis Friedland, *Civic Innovation in America: Community Empowerment, Public*

Policy, and the Movement for Civic Renewal (Berkeley: University of California Press, 2001). See also Elizabeth McKenna, Hahrie Han, and Jeremy Bird, *Groundbreakers: How Obama's 2.2 Million Volunteers Transformed Campaigning in America* (New York: Oxford University Press, 2015).

39. See, for example, Liza Featherstone, "The Vatican Is Way Ahead of the Democratic Party on Climate Change," *New Republic*, June 2, 2023, https://newrepublic.com/article/173187/pope-francis-john-kerry-biden-climate; Tom Wilson and Attracta Mooney, "Church of England Dumps Oil Majors over Climate Concerns," *Financial Times*, June 22, 2023, sec. Anglican Church, https://www.ft.com/content/9af6184a-ed15-4ef4-9c26-d0a9c5c39c1f.
40. Fisher, *Activism, Inc.*, chap. 5.
41. Juheon Lee et al., "Social Capital Building Interventions and Self-Reported Post-Disaster Recovery in Ofunato, Japan," *Scientific Reports* 12, no. 1 (June 17, 2022): 10274, https://doi.org/10.1038/s41598-022-14537-8.
42. Doug McAdam, "Tactical Innovation and the Pace of Insurgency," *American Sociological Review* 48, no. 6 (1983): 735–54, https://doi.org/10.2307/2095322; see also Morris, "Black Southern Student Sit-In Movement."
43. Lara Putnam, Erica Chenoweth, and Jeremy Pressman, "The Floyd Protests Are the Broadest in U.S. History—and Are Spreading to White, Small-Town America," *Washington Post*, June 6, 2020, https://www.washingtonpost.com/politics/2020/06/06/floyd-protests-are-broadest-us-history-are-spreading-white-small-town-america/.
44. For a full discussion, see Dana Fisher, "I'm a Professor Who Studies Protests and Activism: Here's Why the George Floyd Protests Are Different.," Business Insider, June 7, 2020, https://www.businessinsider.com/protests-activism-professor-why-george-floyd-movement-is-different-2020-6.
45. Fisher, interview with Keya Chatterjee, August 11, 2022.
46. James M. Jasper and Jane D. Poulsen, "Recruiting Strangers and Friends: Moral Shocks and Social Networks in Animal Rights and Anti-Nuclear Protests," *Social Problems* 42, no. 4 (1995): 493–512, https://doi.org/10.2307/3097043.
47. Keeanga-Yamahtta Taylor, "Did Last Summer's Black Lives Matter Protests Change Anything?," *New Yorker*, August 6, 2021, https://www.newyorker.com/news/our-columnists/did-last-summers-protests-change-anything.

48. Dana R. Fisher and Stella M. Rouse, "Intersectionality Within the Racial Justice Movement in the Summer of 2020," *Proceedings of the National Academy of Sciences* 119, no. 30 (July 26, 2022): e2118525119, https://doi.org/10.1073/pnas.2118525119.
49. Michael Sainato, "US Sees Union Boom Despite Big Companies' Aggressive Opposition," *Guardian*, July 27, 2022, sec. US news, https://www.theguardian.com/us-news/2022/jul/27/us-union-boom-starbucks-amazon.
50. Greg Rosalsky, "You May Have Heard of the 'Union Boom': The Numbers Tell a Different Story," *NPR*, February 28, 2023, sec. Newsletter, https://www.npr.org/sections/money/2023/02/28/1159663461/you-may-have-heard-of-the-union-boom-the-numbers-tell-a-different-story.
51. Fisher, interview with Jason Walsh, February 4, 2022.
52. Jennifer M. Silva, Sanya Carley, and David M. Konisky, "'I Earned the Right to Build the Next American Car': How Autoworkers and Communities Confront Electric Vehicles," *Energy Research & Social Science* 99 (May 1, 2023): 103065, https://doi.org/10.1016/j.erss.2023.103065.
53. I am not saying we should not continue to push for climate mitigation; but we also must prepare to adapt.
54. United States Environmental Protection Agency, "Using Trees and Vegetation to Reduce Heat Islands," last updated on October 25, 2022, https://www.epa.gov/heatislands/using-trees-and-vegetation-reduce-heat-islands.
55. United States Environmental Protection Agency, "Soak Up the Rain: Trees Help Reduce Runoff," last updated on March 24, 2023, https://www.epa.gov/soakuptherain/soak-rain-trees-help-reduce-runoff.
56. Tom H. Oliver et al., "Empowering Citizen-Led Adaptation to Systemic Climate Change Risks," *Nature Climate Change* 13, (2023): 671–78, https://doi.org/10.1038/s41558-023-01712-6.

METHODOLOGICAL APPENDIX

1. Lorien Jasny, Joseph Waggle, and Dana R. Fisher, "An Empirical Examination of Echo Chambers in US Climate Policy Networks," *Nature Climate Change* 5, no. 8 (August 2015): 782–786, https://doi.org/10.1038/nclimate2666; Lorien Jasny et al., "Shifting Echo Chambers in US Climate Policy Networks," *PLOS ONE* 13, no. 9 (September 14, 2018):

e0203463, https://doi.org/10.1371/journal.pone.0203463; Lorien Jasny and Dana R. Fisher, "Echo Chambers in Climate Science," *Environmental Research Communications* 1, no. 10 (October 11, 2019): 101003, https://doi.org/10.1088/2515-7620/ab491c.

2. Dana R. Fisher, Philip Leifeld, and Yoko Iwaki, "Mapping the Ideological Networks of American Climate Politics," *Climatic Change* 116, no. 3–4 (2013): 523–545; Dana R. Fisher, Joseph Waggle, and Philip Leifeld, "Where Does Political Polarization Come From? Locating Polarization Within the US Climate Change Debate," *American Behavioral Scientist* 57, no. 1 (2013): 70–92; Dana R. Fisher and Philip Leifeld, "The Polycentricity of Climate Policy Blockage," *Climatic Change* 155 (2019): 469–87, https://doi.org/10.1007/s10584-019-02481-y.
3. Open Secrets, "Lobbying Data Summary," accessed October 26, 2021 https://www.opensecrets.org/federal-lobbying/.
4. Because the 116th Congress ended in early January 2021, we did not include 2021 in our sample.
5. United Nations Framework Convention on Climate Change (FCCC), Conference of the Parties, Twenty-first Session, Paris, November 30 to December 11, 2015, "List of Participants," December 11, 2015, http://unfccc.int/resource/docs/2015/cop21/eng/inf03p01.pdf (accessed October 25, 2021).
6. For a discussion of these political moments, see Jasny, Waggle, and Fisher, "An Empirical Examination of Echo Chambers;" Jasny et al., "Shifting Echo Chambers in US Climate Policy Networks;" Jasny and Fisher, "Echo Chambers in Climate Science."
7. Kelly Anne Smith, "Here's What's in the Inflation Reduction Act," Forbes, August 23, 2022, https://www.forbes.com/advisor/personal-finance/inflation-reduction-act/.
8. Jasny, Waggle, and Fisher, "An Empirical Examination of Echo Chambers;" Jasny et al., "Shifting Echo Chambers in US Climate Policy Networks;" Dana R. Fisher et al., "Polarizing Climate Politics in America," in *Environment, Politics, and Society* (Bingley, UK: Emerald Publishing, 2018), 1–23, https://doi.org/10.1108/S0895-993520180000025001; Jasny and Fisher, "Echo Chambers in Climate Science."
9. John Lofland and Lyn H. Lofland, *Analyzing Social Settings: A Guide to Qualitative Observation and Analysis* (Belmont, CA: Wadsworth, 1995).
10. Lofland and Lofland, *Analyzing Social Settings*, 5.
11. Stefaan Walgrave and Joris Verhulst, "Selection and Response Bias in Protest Surveys," *Mobilization: An International Quarterly* 16, no. 2 (June 1, 2011):

203–222, https://doi.org/10.17813/maiq.16.2.j475m8627u4u8177; Stefaan Walgrave, Ruud Wouters, and Pauline Ketelaars, "Response Problems in the Protest Survey Design: Evidence from Fifty-One Protest Events in Seven Countries," *Mobilization: An International Quarterly* 21, no. 1 (March 1, 2016): 83–104, https://doi.org/10.17813/1086/671X-21-1-83.

12. Dana R. Fisher and Stella M. Rouse, "Intersectionality Within the Racial Justice Movement in the Summer of 2020," *Proceedings of the National Academy of Sciences* 119, no. 30 (July 26, 2022): e2118525119, https://doi.org/10.1073/pnas.2118525119.

INDEX

Italicized page numbers indicate figures or tables.

INDEX